W0111290

43 Topics in Current Chemistry
Fortschritte der chemischen Forschung

New Concepts III

Springer-Verlag
Berlin Heidelberg GmbH 1973

This series presents critical reviews of the present position and future trends in modern chemical research. It is addressed to all research and industrial chemists who wish to keep abreast of advances in their subject.

As a rule, contributions are specially commissioned. The editors and publishers will, however, always be pleased to receive suggestions and supplementary information. Papers are accepted for "Topics in Current Chemistry" in either German or English.

Any volume of the series may be purchased separately.

ISBN 978-3-662-15988-0   ISBN 978-3-540-37730-6 (eBook)
DOI 10.1007/978-3-540-37730-6

This work is subject to copyright. All rights are reserved, whether the whole or part of the material is concerned, specifically those of translation, reprinting, re-use of illustrations, broadcasting, reproduction by photocopying machine or similar means, and storage in data banks. Under § 54 of the German Copyright Law where copies are made for other than private use, a fee is payable to the publisher, the amount of the fee to be determined by agreement with the publisher. © by Springer-Verlag Berlin Heidelberg 1973.
Originally published by Springer-Verlag Berlin Heidelberg New York in 1973
Softcover reprint of the hardcover 1st edition 1973

Library of Congress Catalog Card Number 51-5497.

The use of registered names, trademarks, etc. in this publication does not imply, even in the absence of a specific statement, that such names are exempt from the relevant protective laws and regulations and therefore free for general use.

Contents

Editorial Board:

Prof. Dr. *A. Davison*   Department of Chemistry, Massachusetts Institute
                             of Technology, Cambridge, MA 02139, USA

Prof. Dr. *M. J. S. Dewar*   Department of Chemistry, The University of Texas
                             Austin, TX 78712, USA

Prof. Dr. *K. Hafner*   Institut für Organische Chemie der TH
                             D-6100 Darmstadt, Schloßgartenstraße 2

Prof. Dr. *E. Heilbronner*   Physikalisch-Chemisches Institut der Universität
                             CH-4000 Basel, Klingelbergstraße 80

Prof. Dr. *U. Hofmann*   Institut für Anorganische Chemie der Universität
                             D-6900 Heidelberg 1, Im Neuenheimer Feld 7

Prof. Dr. *J. M. Lehn*   Institut de Chimie, Université de Strasbourg, 1, rue
                             Blaise Pascal, B. P. 296/R8, F-67008 Strasbourg-Cedex

Prof. Dr. *K. Niedenzu*   University of Kentucky, College of Arts and Sciences
                             Department of Chemistry, Lexington, KY 40506, USA

Prof. Dr. *Kl. Schäfer*   Institut für Physikalische Chemie der Universität
                             D-6900 Heidelberg 1, Im Neuenheimer Feld 7

Prof. Dr. *G. Wittig*   Institut für Organische Chemie der Universität
                             D-6900 Heidelberg 1, Im Neuenheimer Feld 7

Managing Editor:

Dipl.-Chem. *F. Boschke*   Springer-Verlag, D-6900 Heidelberg 1, Postfach 1780

Springer-Verlag   D-6900 Heidelberg 1 · Postfach 1780
                            Telephone (06221) 49101 · Telex 04-61723

                            D-1000 Berlin 33 · Heidelberger Platz 3
                            Telephone (0311) 822011 · Telex 01-83319

Springer-Verlag   New York, NY 10010 · 175, Fifth Avenue
New York Inc.   Telephone 673-2660

MO Approach to Electronic Spectra of Radicals

Dr. Petr Cársky and Dr. Sc. Rudolf Zahradník

The J. Heyrovský Institute of Physical Chemistry and Electrochemistry, Czechoslovak Academy of Sciences, Prague, Czechoslovakia

Contents

1

P. Čársky and R. Zahradník

I. Introduction

This review concerns the analysis of electronic spectra of radicals by quantum chemical methods. Both semiempirical and ab initio MO methods have been examined, the former in greater detail. The radicals treated are of various structural types, *e.g.* diatomic systems as well as large conjugated molecules. The papers cited do not represent a complete coverage of the literature on this topic, in particular with ab initio calculations. This was dictated by the length of the article and by our desire to present typical applications and representative results of available MO methods. Therefore we have favoured more recent papers and studies whose results appear to be of general importance. Moreover, we are aware that some interesting papers may have escaped our attention.

The optical spectra of radicals provide important data for testing open-shell methods. Until about 1960 open-shell MO studies were rather rare. An intensive development of open-shell MO theory started in the early 'sixties and was followed by chemical applications and systematic studies in the late 'sixties. At present it is possible to state that the physicochemical properties of radicals are predicted with an accuracy comparable to that attained for closed-shell molecules. This is important not only from the viewpoint of the electronic spectra, which can hardly be interpreted without MO theory, but also from the viewpoint of the general theory of reactivity, since radicals and excited states offer the means to overcome spin and symmetry restrictions in certain chemical reactions.

II. Outline of the Open-Shell MO Treatment

A. Choice of SCF Method

With radicals there is no convenient method like the Hartree-Fock-Roothaan procedure commonly used for closed-shell systems. In contrast, the open-shell theory is typical of a number of methods suggested which differ in accuracy from the viewpoint of true SCF theory, in range of applicability, complexity, and computing feasibility. A critical survey of open-shell SCF methods reported by Berthier [1] covers the literature up to 1962. We shall not duplicate that review here; we propose rather to note some features of open-shell methods relevant to their computation feasibility and to mention procedures published after 1962. The unrestricted treatments that assume different space orbitals for different spins will be disregarded here because the restricted wave functions

2

which are eigenfunctions of S^2 and S_z should be preferred in applications to electronic spectra. King et al.[2] used the unrestricted Hartree-Fock CNDO method for interpreting the electronic spectrum of the fluorosulfate radical; a study of this type is, however, rather exceptional.

When discussing open-shell methods, it is convenient to use the formalism of the well-known Roothaan procedure [3]. Hence, consider an open-shell configuration for which the total energy can be expressed as

$$E = 2 \sum_k H_k + \sum_{kl} (2 J_{kl} - K_{kl}) + f \left[2 \sum_m H_m + f \sum_{mn} (2 a J_{mm} - b K_{mn}) + 2 \sum_{km} (2 J_{km} - K_{km}) \right] \tag{1}$$

where all symbols have the conventional meaning [3]. On requiring the vanishing first-order variation of E, the following equations are obtained

$$F_C \varphi_k = \sum_j \varphi_j \vartheta_{jk}, \tag{2}$$

$$F_O \varphi_m = \sum_j \varphi_j \vartheta_{jm}, \tag{3}$$

where

$$F_C = H + 2 J_C - K_C + 2 J_O - K_O, \tag{4}$$

$$F_O = H + 2 J_C - K_C - 2 a J_O - b K_O. \tag{5}$$

As the off-diagonal multipliers, ϑ_{nk}, coupling closed and open shells cannot, in general, be avoided by unitary transformation, a typical problem of the open-shell theory arises here, viz. how to arrive at self-consistency while maintaining the orthogonality of open-shell orbitals to closed-shell orbitals. The simplest way to overcome this drawback is to ignore the ϑ_{nk} multipliers in (2) and (3). This gives the following equations:

$$F_C \varphi_k = \varepsilon_k \varphi_k, \tag{6}$$

$$F_O \varphi_m = \varepsilon_m \varphi_m. \tag{7}$$

This idea was used by Kroto and Santry [4] in CNDO/2 calculations of excited-state properties of HCF, HCN, H_2CO, C_2H_2, and CO_2. The obtained open-shell orbitals in excited singlet configurations were not orthogonal to the doubly occupied orbitals, as expected, but the resulting deviations of the eigenvectors from orthogonality, as measured by the appropriate scalar products, were found to be of the order of 0.02 to

3

0.05 and in the poorest case, with CO_2, 0.15. Hence it appears that this approximate open-shell procedure might be useful in semiempirical MO treatments of radicals. It is noteworthy that Eqs. (6) and (7) are entirely justifiable with configurations where ϑ_{nk} multipliers vanish on symmetry grounds, e.g. with the ground-state configuration of planar methyl [5]. In general, the orthogonality between open and closed shells can be preserved in SCF calculations in several ways by using:

    a) coupling operators
    b) orthogonality-constrained basis set expansion (OCBSE)
    c) the density-matrix method
    d) effective Hamiltonians.

Coupling operators, introduced first by Roothaan [3], enter the Hartree-Fock operator as additional terms to operators determining closed-shell and open-shell orbitals, in the particular case mentioned above to F_C and F_O operators (Eqs. (4) and (5)). The coupling operators are so constructed as to fulfil the two following requirements. Firstly, they involve off-diagonal ϑ_{nk} multipliers, ensuring orthogonality of open-shell orbitals to closed-shell orbitals. Secondly, when added to F_C and F_O, the operators determining closed- and open-shell orbitals take the same form of F, so that both open- and closed-shell orbitals can now be calculated by a single eigenvalue problem

$$F \varphi_i = \varepsilon_i \varphi_i \tag{8}$$

The original Roothaan procedure [3] is applicable to configurations whose total energy can be expressed by the general formula (1). With molecules, the following types of states satisfy this limitation: configurations with one nondegenerate singly occupied MO; configurations of highly symmetrical systems with a degenerate open-shell MO; and configurations with more than one nondegenerate singly occupied MO having all spins parallel (half-closed shell). The LCAO expansion in the ZDO approximation was reported by Adams and Lykos [6], the incorporation into the CNDO computational scheme has been described in a recent review [7]. Huzinaga [8,9] extended the applicability of the Roothaan procedure to other important types of molecular states. His derivation of the eigenvalue problem proceeds along the same lines as in Roothaan's theory, but it starts with a more general energy expression than Eq. (1). Besides the three examples mentioned above, Huzinaga's method also can handle electronic configurations with two open shells of different symmetry, e.g. the $(\pi_u)^M (\pi_g)^N$ configuration. Finally, Birss and Fraga [10] constructed coupling operators of a new form which permitted SCF orbitals to be obtained for electronic systems without any restriction on the number or symmetry of open-shell orbitals.

This advanced state of the open-shell theory looked promising for MO applications in radical chemistry. Indeed, the Roothaan procedure soon became popular and has been employed in various chemical applications. The actual calculations, however, have often encountered convergence difficulties. In general, the Roothaan method is rather slow to converge, sometimes even diverging [11-13]. This difficulty will probably also arise with the methods of Huzinaga and of Birss and Fraga, though here no numerical calculations have yet been reported.

The second family of related methods we characterize by the concept of the orthogonality-constrained basis set expansion (OCBSE), introduced by Hunt and coworkers [14]. In methods of this type the orthogonality conditions between orbitals are not preserved by adding them to the energy variation as constraints. In fact, one employs simple operators such as F_C and F_O, maintaining the orthogonality between orbitals by the device of changing the basis set. For the sake of simplicity, let us consider a case which is amenable to the original Roothaan procedure as given by Eqs. (1)—(5). Let n be the dimension of the basis set and u the number of closed-shell orbitals generated by F_C. If open- and closed-shell orbitals are to be mutually orthogonal, then the former can have no projection on the latter and vice versa, $i.e.$ the MOs of one set can have no admixture of those of the other set. Since Eq. (3) is invariant to any orthogonal transformation, we may change the variational basis set for the solution of Eq. (3) to the $n—u$ dimensional set spanned by the virtual orbitals of F_C, which is called by Hunt $et\ al.$ [14] the "orthogonality-constrained basis set expansion". In actual calculations one proceeds as follows [15]:

a) Construct the F_C and F_O operators in terms of the AO basis set
b) Solve Eq. (6) as a standard eigenvalue problem
c) Construct the F_O matrix over AOs and transform it by the matrix constructed from the $n—u$ virtual orbitals of F_C arranged as column vectors, C_t,

$$F_O' = C_t^+ F_O C_t , \qquad (9)$$

where the F_O' and F_O matrices are of dimension $(n-u) \times (n-u)$ and $n \times n$, respectively, and C_t^+ and C_t of $(n-u) \times n$ and $n \times (n-u)$, respectively.

d) Vectors C_O' resulting as solutions of the Eq. (10)

$$F_O' C_O' = C_O' \varepsilon_O' \qquad (10)$$

may be expanded into the original basis set by

$$C_{Ol\mu} = \sum_{j=1}^{n-u} C_{Oj\mu}' C_{lj} \qquad (11)$$

5

e) The procedure is iterated back to a) until self-consistency is achieved.

Methods of this type, as developed by Hunt et al. [14] and Segal [15], are applicable to any problem, regardless of the symmetry of the state under consideration. A similar projection technique was developed by Chang et al. [16] for treating configurations with one nondegenerate open-shell orbital. It should be stressed that the prototype for the methods just mentioned was the "second" method of Huzinaga [9]. In that method one uses as many effective F operators as there are closed and open shells (by the shell we imply a single orbital if it is not degenerate). The respective F operators are obtained by partial variations of the total energy expression with respect to individual orbitals. Each shell is optimized through the particular eigenvalue problem of the type of Eq. (10) while the other orbitals are held fixed. The orthogonality between non-fixed and fixed orbitals is preserved by the OCBSE technique. The MOs are optimized stepwise in an iterative process until self-consistency is achieved. Reported calculations [15,16] showed that SCF procedures of the OCBSE type are fairly well convergent.

The density-matrix method of McWeeny [1,17] was found to converge in all cases, but too slowly for practical purposes. The energy-weighted steepest descent (EWSD) method [18], developed by Hillier and Saunders, is related to the McWeeny method; it is claimed [18] that the modifications adopted bring about improved convergence.

The last group in our classification comprises two approximate SCF procedures which give wave functions that are not correct to first order. The first of them, Nesbet's method of symmetry and equivalence restrictions [19], uses the Hamiltonian of the unrestricted method for the α-spin electrons, the number of α-spin electrons being greater than that of β-spin electrons. The β-spin electrons are forced to occupy MOs given for α-spin electrons by [1]

$$(H + \sum_{j=1,2..} J_j - \sum_{j \text{ odd}} K_j) \, \varphi_i = \varepsilon_i \, \varphi_i . \tag{12}$$

The second method of this group, the method of Longuet-Higgins and Pople [20], with its extension for accommodating states with degenerate open shells is described in detail in the next section.

The two approximate SCF methods just mentioned are advantageous because of their simplicity and rapid convergence. They have proved useful with both semiempirical and ab initio treatments (vide infra). Their approximate nature is not associated, at least in semiempirical calculations, with any drawback, inasmuch as we found in PPP-like and CNDO treatments [13,21,22] that the methods of Roothaan and of Longuet-Higgins and Pople, combined with a limited configuration inter-action, gave transition energies that were very close in absolute value. With ab initio calculations, however, the use of these methods is some-what delicate because a departure from the true SCF theory means a certain "degradation" of nonempirical calculations. As the poor conver-gence rules out procedures using coupling operators and the density-matrix method, methods of the OCBSE type appear to be the most convenient SCF procedures for ab initio calculations.

There are several MO approaches to electronic spectra at different levels of sophistication. The simplest of them makes use of the SCF excitation energies based on the virtual orbital approximation. In a more advanced approach, both ground- and excited-state configurations are subjected to variational SCF treatments, the respective energy differences being the predicted transition energies. In spite of the suc-cess of SCF treatments in some cases, SCF calculations should, in general, be followed by configuration interaction (CI). With semiempi-rical SCF-CI calculations, the virtual orbital approximation is ordinarily used. In ab initio calculations the iterative natural-orbital method of Bender and Davidson [23,24] has proved very useful (cf. e.g. [25,26]).

B. Choice of the Configuration Interaction Basis

We shall mention here only the situation with semiempirical methods. Comparative studies [27,28] within the π-electron approximation have demonstrated that the doubly excited states had little effect upon the predicted electronic spectra. Amos and Woodward [29] reported complete CI calculations of the PPP-type on allyl, pentadienyl, and the mono-positive and mononegative butadiene radical ions. We calculated [30,31] the same systems using only the singly excited configurations. The resulting differences in transition energies do not exceed 2000 cm^{-1} and are generally less than 1000 cm^{-1}. Hence, it is sufficient to employ a CI basis covering configurations which correspond to one-electron promotions among the several highest occupied MOs, singly occupied MOs, and several lowest MOs unoccupied in the ground state. This also probably applies to semiempirical all-valence electron methods, inasmuch as Brabant and Salahub [32] found with closed-shell molecules that the doubly excited states had very little effect on the calculated spectral data. Finally, it should be pointed out that the CI treatment need not be based on MOs given by the open-shell SCF procedure. The ground- and excited-state configurations of a radical can be constructed from the MOs of a parent closed-shell molecule, e.g. for the allyl radical one can use the MOs of the allyl cation [33]. The treatment using a true SCF open-shell theory is, however, to be preferred.

7

III. Description of Semiempirical SCF-CI Calculations Using the Method of Longuet-Higgins and Pople

The SCF method of Longuet-Higgins and Pople [20] (LHP) became rather popular in semiempirical open-shell treatments and the majority of the reported semiempirical open-shell calculations have been based on it, thus we consider it expedient to describe it in detail in this section. Longuet-Higgins and Pople formulated [20] the following effective Hamiltonian

$$F = H + \sum_k (2J_k - K_k) + f \sum_m (2J_m - K_m), \tag{13}$$

which generates doubly occupied, singly occupied, and virtual MOs. Here k and m are indices for closed-shell and open-shell orbitals, respectively, and f is the fractional occupation of the open shell as in Eq. (1). Originally, the LHP method was developed for configurations with one nondegenerate singly occupied MO, i.e. for one φ_m and $f = \frac{1}{2}$. Dewar et al. [34] derived this procedure in a different way, called it "the half-electron method" and extended it to treatments of triplet states [35]. Recently a generalization of this method was reported [36] which makes it possible to treat the following configurations by means of the F operator (as given by Eq. (13)): doublet states having one nondegenerate, singly occupied MO; triplet and singlet states with two open-shell MOs that are nondegenerate or form a doubly degenerate open shell; and doublet states with a doubly degenerate open-shell MO occupied by one or three electrons.

In the ZDO approximation, the LHP method is well-adapted for computer programming. In fact, one can use any program for closed-shell molecules by making two small modifications: firstly, the electron-density and bond-order matrix is defined for radicals as follows

$$P_{\mu\nu} = \sum_k 2c_{k\mu} c_{k\nu} + \sum_m 2f c_{m\mu} c_{m\nu}, \tag{14}$$

where the second term accounts for the presence of open shells; secondly, the expression for the total electronic energy must be modified as follows [36]

$$E = E_R + a J_{mm} + b J_{nn} + c J_{mn} + d K_{mn}. \tag{15}$$

Here E_R stands for the term which can be called the "Roothaan" energy because its form in the LCAO expansion

8

$$E_{\text{R}} = \sum_k \varepsilon_k + \sum_m f\,\varepsilon_m + \frac{1}{2} \sum_\mu \sum_\nu H_{\mu\nu}\,P_{\mu\nu} \tag{16}$$

closely resembles the standard closed-shell procedure. Here ε_k and ε_m are orbital energies of closed and open shells given by the F operator (Eq. (13)) and $P_{\mu\nu}$ are evaluated through Eq. (14). Hence, one can arrive at open-shell results using a closed-shell program by changing the P matrix and adding some corrections to the total energy expression. The constants a, b, c, d of Eq. (15), depending on the specific case, are summarized in Table 1. We found that the SCF calculations were rapidly convergent [13] for all types of configurations considered in Table 1.

Table 1. Constants in the total energy expression [36] (15)

Configuration	Spin	f	a	b	c	d
φ_m	Doublet	$1/2$	$-1/4$	0	0	0
$\varphi_m \pm \varphi_n$	Doublet	$1/4$	$-1/16$	$-1/16$	$-1/4$	$1/8$
$\varphi_m\,\varphi_n$	Triplet	$1/2$	$-1/4$	$-1/4$	0	$-1/2$
$\varphi_m\,\varphi_n$	Singlet	$1/2$	$-1/4$	$-1/4$	0	$3/2$
$\varphi_m^2 - \varphi_n^2$	Singlet	$1/2$	$1/4$	$1/4$	-1	$-1/2$
$\varphi_m^2 + \varphi_n^2$	Singlet	$1/2$	$1/4$	$1/4$	-1	$3/2$
$\varphi_m^2\,\varphi_n \pm \varphi_m\,\varphi_n^2$	Doublet	$3/4$	$-1/16$	$-1/16$	$-1/4$	$1/8$

As stated in the preceding section, it is sufficient to consider only singly excited states in the CI treatment. For the simplest case, where the ground-state configuration has one nondegenerate singly occupied MO, the types of singly excited states are presented in Fig. 1 and the respective wave functions can be expressed as follows

$$^2\Psi_{\text{G}} = |\varphi_1\,\bar{\varphi}_1 \cdots\cdots\cdots\cdots \varphi_{m-1}\,\bar{\varphi}_{m-1}\,\varphi_m| \tag{17}$$

$$^2\Psi_{\text{A}} = |\varphi_1\,\bar{\varphi}_1 \cdots\cdots \varphi_i\,\bar{\varphi}_m \cdots\cdots \varphi_{m-1}\,\bar{\varphi}_{m-1}\,\varphi_m| \tag{18}$$

$$^2\Psi_{\text{B}} = |\varphi_1\,\bar{\varphi}_1 \cdots\cdots\cdots\cdots \varphi_{m-1}\,\bar{\varphi}_{m-1}\,\varphi_k| \tag{19}$$

$$^2\Psi_{\text{C}a} = 1/\sqrt{2}(|\varphi_1\,\bar{\varphi}_1 \cdots\cdots \varphi_i\,\bar{\varphi}_k \cdots\cdots \varphi_{m-1}\,\bar{\varphi}_{m-1}\,\varphi_m|$$
$$+ |\varphi_1\,\bar{\varphi}_1 \cdots\cdots \varphi_k\,\bar{\varphi}_i \cdots\cdots \varphi_{m-1}\,\bar{\varphi}_{m-1}\,\varphi_m|) \tag{20}$$

9

P. Čársky and R. Zahradník

$$^2\Psi_{C_\beta} = 1/\sqrt{6}(\ |\varphi_1\,\bar{\varphi}_1 \cdots\cdots \varphi_i\,\bar{\varphi}_k \cdots\cdots \varphi_{m-1}\,\bar{\varphi}_{m-1}\,\varphi_m|$$
$$-\ |\varphi_1\,\bar{\varphi}_1 \cdots\cdots \varphi_k\,\bar{\varphi}_i \cdots\cdots \varphi_{m-1}\,\bar{\varphi}_{m-1}\,\varphi_m| \qquad (21)$$
$$+\ 2\ |\varphi_1\,\bar{\varphi}_1 \cdots\cdots \varphi_i\,\bar{\varphi}_m \cdots\cdots \varphi_{m-1}\,\bar{\varphi}_{m-1}\,\varphi_k\ |)$$

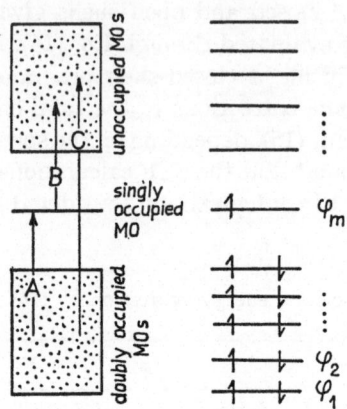

Fig. 1. Types of electron promotions leading to singly excited states. Configurations C_α and C_β differ by the spin function (see text)

The CI matrix elements based on the LHP MOs have been reported by Ishitani and Nagakura [37)]

$$<{}^2\Psi_A(i \longrightarrow m)\,|H|{}^2\Psi_A(i \longrightarrow m)> - <{}^2\Psi_G|H|{}^2\Psi_G> =$$
$$= \varepsilon_m - \varepsilon_i + \frac{1}{2}\,[(im\,|G|\,mi) + (mm\,|G|\,mm) - 2\,(im\,|G|\,im)] \quad (22)$$

$$<{}^2\Psi_B(m \longrightarrow k)\,|H|{}^2\Psi_B(m \longrightarrow k)> - <{}^2\Psi_G\,|H|{}^2\Psi_G> =$$
$$= \varepsilon_k - \varepsilon_m + \frac{1}{2}\,[(mk\,|G|\,km) + (mm\,|G|\,mm) - 2\,(mk\,|G|\,mk)] \quad (23)$$

$$<{}^2\Psi_{C_\alpha}(i \longrightarrow k)\,|H|{}^2\Psi_{C_\alpha}(i \longrightarrow k)> - <{}^2\Psi_G\,|H|{}^2\Psi_G> =$$
$$= \varepsilon_k - \varepsilon_i + 2\,(ik\,|G|\,ki) - (ik\,|G|\,ik) \qquad (24)$$

$$<{}^2\Psi_{C_\beta}(i \longrightarrow k)\,|H|{}^2\Psi_{C_\beta}(i \longrightarrow k)> - <{}^2\Psi_G\,|H|{}^2\Psi_G> =$$
$$= \varepsilon_k - \varepsilon_i + (im\,|G|\,mi) + (mk\,|G|\,km) - (ik\,|G|\,ik) \qquad (25)$$

$$<{}^2\Psi_G\,|H|{}^2\Psi_A(i \longrightarrow m)> = \frac{1}{2}\,(im\,|G|\,mm) \qquad (26)$$

10

$$<{}^2\Psi_{\mathrm{G}}|H|{}^2\Psi_{\mathrm{B}}(m \longrightarrow k)> = -\frac{1}{2}(mm|G|mk) \tag{27}$$

$$<{}^2\Psi_{\mathrm{G}}|H|{}^2\Psi_{\mathrm{C}_\alpha}(i \longrightarrow k)> = 0 \tag{28}$$

$$<{}^2\Psi_{\mathrm{G}}|H|{}^2\Psi_{\mathrm{C}_\beta}(i \longrightarrow k)> = \frac{\sqrt{6}}{2}(im|G|mk) \tag{29}$$

$$<{}^2\Psi_{\mathrm{A}}(i \longrightarrow m)|H|{}^2\Psi_{\mathrm{A}}(h \longrightarrow m)> = \frac{1}{2}(mh|G|im) - \\ - (mh|G|mi) \ (h \neq i) \tag{30}$$

$$<{}^2\Psi_{\mathrm{A}}(i \longrightarrow m)|H|{}^2\Psi_{\mathrm{B}}(m \longrightarrow k)> = (im|G|mk) \tag{31}$$

$$<{}^2\Psi_{\mathrm{A}}(h \longrightarrow m)|H|{}^2\Psi_{\mathrm{C}_\alpha}(i \longrightarrow k)> = \frac{\sqrt{2}}{2}[2\,(im|G|kh) - \\ - (im|G|hk) + \delta_{hi}\frac{1}{2}(mm|G|mk)] \tag{32}$$

$$<{}^2\Psi_{\mathrm{A}}(h \longrightarrow m)|H|{}^2\Psi_{\mathrm{C}_\beta}(i \longrightarrow k)> = \frac{\sqrt{6}}{2}[\delta_{hi}\frac{1}{2}(mm|G|mk) - \\ - (im|G|hk)] \tag{33}$$

$$<{}^2\Psi_{\mathrm{B}}(m \longrightarrow k)|H|{}^2\Psi_{\mathrm{B}}(m \longrightarrow l)> = \frac{1}{2}\,(mk|G|lm) - \\ - (mk|G|ml) \ (k \neq l) \tag{34}$$

$$<{}^2\Psi_{\mathrm{B}}(m \longrightarrow l)|H|{}^2\Psi_{\mathrm{C}_\alpha}(i \longrightarrow k)> = \frac{\sqrt{2}}{2}[2\,(il|G|km) - \\ - (il|G|mk) + \delta_{kl}\frac{1}{2}(im|G|mm)] \tag{35}$$

$$<{}^2\Psi_{\mathrm{B}}(m \longrightarrow l)|H|{}^2\Psi_{\mathrm{C}_\beta}(i \longrightarrow k)> = \frac{\sqrt{6}}{2}[(il|G|mk) - \\ - \delta_{kl}\frac{1}{2}(im|G|mm)] \tag{36}$$

$$<{}^2\Psi_{\mathrm{C}_\alpha}(i \longrightarrow k)|H|{}^2\Psi_{\mathrm{C}_\alpha}(h \longrightarrow l)> = 2\,(hk|G|li) - \\ - (hk|G|il) \ (k \neq l) \ \text{or} \ (i \neq h) \tag{37}$$

$$<{}^2\Psi_{\mathrm{C}_\alpha}(i \longrightarrow k)|H|{}^2\Psi_{\mathrm{C}_\beta}(h \longrightarrow l)> = \frac{\sqrt{3}}{2}[\delta_{hi}(km|G|ml) - \\ - \delta_{kl}(hm|G|mi)] \tag{38}$$

P. Čársky and R. Zahradník

$$<^2\Psi_{C_\beta}(i \longrightarrow k)|H|^2\Psi_{C_\beta}(h \longrightarrow l)> = \delta_{hi}(mk|G|lm) +$$
$$+ \delta_{kl}(hm|G|mi) - (hk|G|il) \tag{39}$$
$$(h \neq i) \text{ or } (k \neq l)$$

In Eqs. (22)—(39) ε_k, ε_m, and ε_l mean LHP orbital energies, h and i are indices for doubly occupied MOs, m for the singly occupied MO, k and l for virtual MOs, and the notation for repulsion integrals is as follows

$$(hk|G|il) = \iint \varphi_h(1)\, \varphi_k(2)\, \frac{e^2}{r_{12}}\, \varphi_i(1)\, \varphi_l(2)\, d\tau \tag{40}$$

Expressions for $<^2\Psi_1|\vec{r}|^2\Psi_2>$, which are necessary for calculations of oscillator strengths, are summarized in Table 2. For the sake of completeness, let us mention the CI treatment of quartet states. The

Table 2. Transition moments among doublet [37] (G, A, B, C_α, C_β) and quartet [7] (Q) configurations

	Transition moments $r_{ij} = \int \varphi_i\, r\varphi_j\, d\tau$		Transition moments $r_{ij} = \int \varphi_i\, r\varphi_j\, d\tau$
G—A($i \to m$)	r_{im}	B($m \to l$)—$C_\beta(i \to k)$	0
G—B($m \to k$)	r_{mk}	$C_\alpha(i \to k)$—$C_\alpha(i \to l)$	r_{kl}
G—$C_\alpha(i \to k)$	$\sqrt{2}\, r_{ik}$	$C_\alpha(i \to k)$—$C_\alpha(h \to k)$	$-r_{hi}$
G—$C_\beta(i \to k)$	0	$C_\alpha(i \to k)$—$C_\alpha(h \to l)$	0
A($i \to m$)—A($h \to m$)	$-r_{hi}$	$C_\alpha(i \to k)$—$C_\beta(i \to k)$	0
A($i \to m$)—B($m \to k$)	0	$C_\alpha(i \to k)$—$C_\beta(i \to l)$	0
A($i \to m$)—$C_\alpha(i \to k)$	$(1/\sqrt{2})\, r_{mk}$	$C_\alpha(i \to k)$—$C_\beta(h \to k)$	0
A($h \to m$)—$C_\alpha(i \to k)$	0	$C_\alpha(i \to k)$—$C_\beta(h \to l)$	0
A($i \to m$)—$C_\beta(i \to k)$	$(\sqrt{6}/2)\, r_{mk}$	$C_\beta(i \to k)$—$C_\beta(i \to l)$	r_{kl}
A($h \to m$)—$C_\beta(i \to k)$	0	$C_\beta(i \to k)$—$C_\beta(h \to k)$	$-r_{hi}$
B($m \to k$)—B($m \to l$)	r_{kl}	$C_\beta(i \to k)$—$C_\beta(h \to l)$	0
B($m \to k$)—$C_\alpha(i \to k)$	$(-1/\sqrt{2})\, r_{im}$	Q($i \to k$)—Q($i \to l$)	r_{kl}
B($m \to l$)—$C_\alpha(i \to k)$	0	Q($i \to k$)—Q($h \to k$)	$-r_{hi}$
B($m \to k$)—$C_\beta(i \to k)$	$(\sqrt{6}/2)\, r_{im}$	Q($i \to k$)—Q($h \to l$)	0

wave function for quartets is

$$^4\Psi = \frac{1}{\sqrt{3}} (|\varphi_1\, \bar{\varphi}_1 \cdots \varphi_i\, \varphi_k \cdots \bar{\varphi}_m|$$
$$+ |\varphi_1\, \bar{\varphi}_1 \cdots \varphi_i\, \bar{\varphi}_k \cdots \varphi_m| \tag{41}$$
$$+ |\varphi_1\, \bar{\varphi}_1 \cdots \bar{\varphi}_i\, \varphi_k \cdots \varphi_m|)$$

12

and the CI matrix elements based on the LHP MOs for the ground doublet state are

$$<{}^4\Psi(i \longrightarrow k)|H|{}^4\Psi(i \longrightarrow k)> - <{}^2\Psi_G|H|{}^2\Psi_G> =$$

$$= \varepsilon_k - \varepsilon_i - (ik|G|ik) - \frac{1}{2}(mk|G|km) - \frac{1}{2}(mi|G|im) \qquad (42)$$

$$<{}^4\Psi(i \longrightarrow k)|H|{}^4\Psi(i \longrightarrow l)> = -(ik|G|il) - \frac{1}{2}(km|G|ml) \qquad (43)$$

$$<{}^4\Psi(i \longrightarrow k)|H|{}^4\Psi(h \longrightarrow k)> = -(hk|G|ik) - \frac{1}{2}(mh|G|im) \qquad (44)$$

$$<{}^4\Psi(i \longrightarrow k)|H|{}^4\Psi(h \longrightarrow l)> = -(hk|G|il) \qquad (45)$$

The $<\Psi_1|\vec{r}|\Psi_2>$ expressions for quartets are given in Table 2. Finally, it is necessary to specify the additional details of the calculations whose results are discussed in Section V. If not otherwise noted there, the following specification applies. The π-electron calculations are of te PPP type, using the Mataga and Nishimoto formula for repulsion integrals. Idealized geometries were used throughout the calculations: all trigonal C—C bond lengths are 1.40 Å and bond angles in polyenic chains are 120°. For valence-state ionization potentials, one-center repulsion integrals, and resonance integrals, we use the following parameters (in eV): $I_C = 11.22$; $I_N = 14.1$; $\gamma_{CC} = 10.53$; $\gamma_{NN} = 12.3$; $\beta_{CN} = \beta_{CC} = -2.318$. All approximations and semiempirical parameters adopted in the CNDO calculations are due to Del Bene and Jaffé [38]. However, we use the Mataga-Nishimoto approximation for two-center repulsion integrals rather than that of Pariser and Parr (in a later paper[39] Jaffé and coworkers also used this modification). We extended the computational scheme of Del Bene and Jaffé to make it applicable to systems containing boron and fluorine; from the trends in CNDO/2 parameters for hydrogen, carbon, nitrogen, and oxygen, we tentatively chose the following parameters: $\gamma_{BB} = 10.2$ eV; $\beta_B^o = -15$; $\gamma_{FF} = 13.9$ eV; $\beta_F^o = -55$. Wherever available we used the experimental geometries, otherwise they were inferred from similar molecules (cf. [22,40]).

In both PPP-like and CNDO calculations, the configuration interaction treatment has been based on the virtual MO approximation, considering all configurations arising formally from one-electron transitions between several highest doubly occupied orbitals, the singly occupied orbital, and several lowest vacant orbitals.

13

IV. Source of Experimental Data and Their Presentation

To our knowledge there is no collection of absorption curves comprising the electronic spectra of radicals of various structural types. The comprehensive collection of Habersbergerová *et al.* [41] summarizes in the tabulated form the electronic spectral data for various radicals generated by irradiation techniques. Many references to papers on electronic spectra of aromatic radicals can be found in the review of Land [42]. The results of Hamill and co-workers are summarized in the review [43] which presents many absorption curves of radical ions (predominantly of aromatic hydrocarbons) produced in rigid glasses by γ-irradiation. Data on the electronic spectra of small radicals are provided by the books of Herzberg [44,45]. Finally it should be pointed out that a great deal of information on the electronic spectra of radical cations can be inferred from the photoelectron spectral data (vide infra).

The majority of the experimental data treated in the next section have been taken from the papers of Hoijtink and his co-workers [46-49] and from Herzberg [44,45]. The absorption curves in the original papers were redrawn using the $\log \varepsilon$ and $\tilde{\nu}(k\mathrm{K})$ scales, their ratio being $1:10$ as used in collections of spectral data (*e.g.* [50]). The allowed electronic transitions, predicted by the LCI–SCF calculations, have been entered in the figures with absorption curves as vertical lines. Using the empirical relationship $\log \varepsilon = \log f + 4.5$, the heights of the lines represent the predicted intensities (f means the predicted oscillator strength). Predicted forbidden transitions are indicated by wavy lines with arrows. Experimental transition energies tabulated or given in the text refer to vertical transitions; if they were not explicitly given in the original publications, we read them at the positions of the highest absorption maxima. In some cases, in particular with small radicals, only the regions of the observed absorption are available. The transition energies in radical cations can be obtained [51] from the photoelectron spectral data in the way depicted in Fig. 2, *i.e.* by subtracting the first ionization potential from the higher ones. In this case, however, the open-shell semiempirical calculations should be performed for the geometry of the parent closed-shell system.

V. Survey of Applications

For the sake of convenience we classify the MO studies from the viewpoint of the MO method used rather than of the structure of radicals treated. Accordingly, we present separately the results of open-shell PPP-like, semiempirical all-valence electron, and ab initio calculations.

14

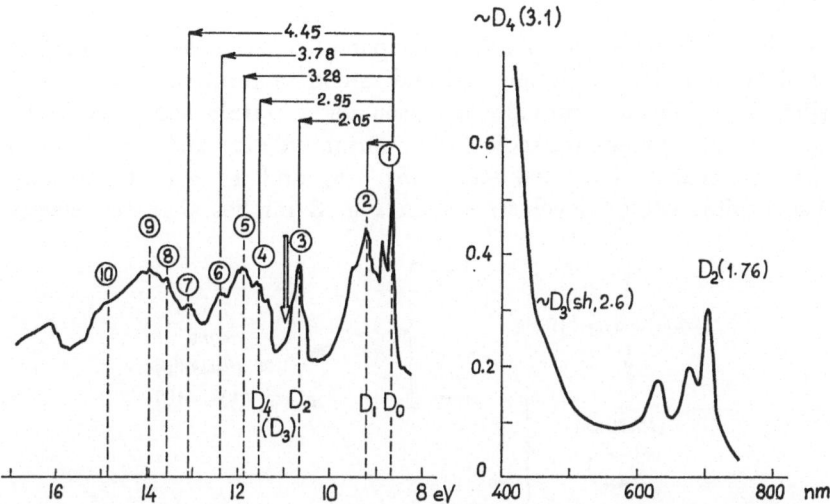

Fig. 2. Determination of transition energies in the quinoline radical cation from the photoelectron spectrum of quinoline [52]. Presence of a hidden peak (⇒) in the photoelectron spectrum (left) and of the shoulder (D_3) observed in the electronic spectrum [53] of the quinoline radical cation is assumed on the basis of semiempirical calculations [51] (right: the values in parentheses are in eV)

For the chemical classification we attempted to work solely with the π-electron calculations, which cover the conjugated radicals. Basic information on calculations and experimental data is given in the preceding sections; for additional details, see the original publications. Before presenting a confrontation of theoretical and experimental data, we note some general features in the nature of electronic spectra of radicals.

A. Common Features in Electronic Spectra of Open-Shell Systems

Electronic spectra of radicals differ from those of closed-shell molecules in several points. We mention here the most outstanding of them.

i) The first transition energies have been found to fall into the visible, near-infrared, or even into the infrared regions, not only with large conjugated radicals but also with systems as small as those containing five or slightly more valence electrons. The NH^+ radical represents the extreme case, its first electronic transition ($^4\Sigma^- \leftarrow {}^2\Pi$) being located at about 320 cm^{-1}. With BH_2, which has the same number of electrons, absorption starts at 11,600 cm^{-1} and the first band of the three-valence electron BeH is located at 20,000 cm^{-1}. Numerous small conjugated

P. Čársky and R. Zahradník

radicals also absorb in the near-infrared region. In general, a remarkable red shift is observed on going from the parent closed molecule to an open-shell system. This shift may be understood in terms of HMO theory (Fig. 3), which also provides a reason for a certain similarity in the spectra of mononegative radical ions, dinegative ions, and parent closed-shell molecules in excited states ($S_1 \rightarrow S_x$ and $T_1 \rightarrow T_x$ transitions). The orbital energy level scheme in Fig. 3 implies that the energy-

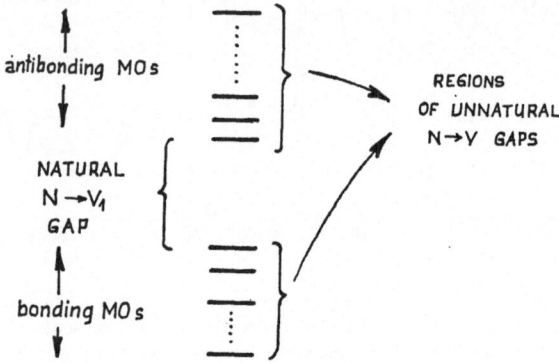

Fig. 3. Typical scheme of MO energy levels of a hydrocarbon

poorest electronic transitions in radical ions are associated with what we call the "unnatural" $N \rightarrow V$ gaps. Ordinarily, the $N \rightarrow V_1$ gap of the parent system is considerably larger than the unnatural gaps, as de-

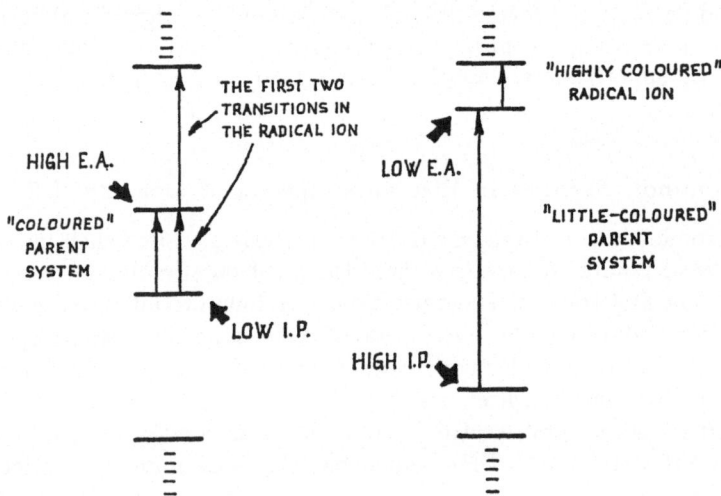

Fig. 4. MO energy scheme of the "coloured" and the "little-coloured" parent systems

16

picted in Fig. 4. The two typical orbital level schemes refer to the "little-coloured" and "highly-coloured" parent systems. The former indicates a considerable shift on going to a radical ion whereas with the latter a rather small shift is to be expected.

ii) In contrast to closed-shell molecules, for the overwhelming majority of radicals the lowest excited state is not of higher spin multiplicity with respect to the ground state, *i. e.* the lowest excited state is a doublet and not a quartet state. This is a rather unfavourable situation for a population of the quartet state and consequently also for the observation of phosphorescence. Accordingly, the decacyclene mononegative ion is the sole aromatic radical for which phosphorescence has been observed [54]. A somewhat striking location of the lowest quartet state above one or several next excited doublet states can be explained by means of the definition of singly excited states in Fig. 1. The lowest-energy electronic transition and usually also the next several ones are of the A and B types. As the formation of a quartet configuration requires three open-shell MOs, the A and B states can only be the doublet states. The C-type transitions lead to both doublet and quartet states, the latter lying lower than the former but almost always still higher than the lowest A-type or B-type state. One exception (NH^+) has already been mentioned in paragraph (i); another is the O_2^+ radical whose $^4\Pi_u \leftarrow {}^2\Pi_g$ transition is located at 31,900 cm^{-1}. The wavelength of the latter gives reason to hope that the phosphorescence might be detected. Both NH^+ and O_2^+ possess ground-state configurations with a degenerate π-molecular orbital occupied by a single electron, which is the case where the lowest C-type configuration in the HMO model is not disfavoured on energy grounds.

iii) In contrast to closed-shell systems, the dependence of the first transition energy on the size of conjugation need not be monotonous within a class of related compounds: sometimes the passage to a larger system is associated with a hypsochromic shift, *e.g.* on going from naphthalene$^-$ to anthracene$^-$ or from diphenyl$^-$ to stilbene$^-$. This is due to the fact that the first band can be assigned to the transition in the unnatural N $\to$ V gap (Fig. 3).

B. Large and Medium-Sized Conjugated Radicals

We introduce here for the conjugated hydrocarbon radicals the same classification which we used for the closed-shell hydrocarbons [55]. We divide the hydrocarbons into two large groups: alternant and non-alternant. Further classification concerns the even and odd systems, and the presence of cycles in the skeleton. The phenyl substituents are not considered to be part of the skeleton, therefore we treat *e.g.* 1,4-diphenylbutadiene as a system "without a cycle".

1. Alternant Even Hydrocarbons

The systems without a cycle are represented here by the series of radical ions derived from polyenes (*1—5*) and α,ω-diphenylpolyenes (*6—10*). With the former, the absorption curves are only available for the butadiene radical ions [56,57], whereas with the latter the situation is

$$CH_2=CH-[CH=CH]_n\,H$$

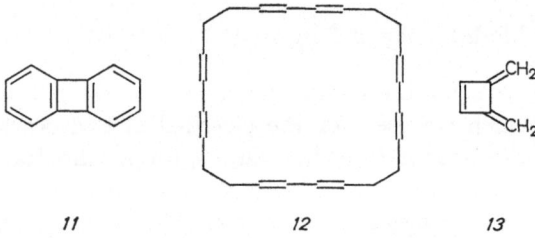

1 (n = 1)	*6 (n = 1)*
2 (n = 2)	*7 (n = 2)*
3 (n = 3)	*8 (n = 3)*
4 (n = 4)	*9 (n = 4)*
5 (n = 5)	*10 (n = 5)*

more satisfactory: the absorption curves for the radical anions derived from *6—10* have been published by Hoijtink *et al.*[47] and for radical cations of *6—8* by Shida and Hamill [43,58]. The available experimental data [47,56] for radical anions are compared with the results of PPP-like semiempirical calculations [30,59] in Figs. 5 and 6. It can be seen that the theory reproduces well the somewhat strange nature of the electronic spectra of the radical anions of *6—10*, *i.e.* strong absorption at 10—20 *k*K followed by a broad region of weak absorption from 20—30 to 45—50 *k*K. We believe that the predicted spectral data for higher polyene radical anions are also reasonable because they are in remarkable agreement with the results of nonempirical π-electron calculations [60]. Long-wavelength absorption of the γ-irradiated butadiene [56,57] has been assigned [57] to the dimeric butadiene radical cation (M_2^+). The CNDO-type calculation [61] (Fig. 7) supports this assignment.

Among nonbenzenoid cyclic systems we mention radical ions derived from hydrocarbons *11—13*. With diphenylene (the dibenzo derivative of

11   *12*   *13*

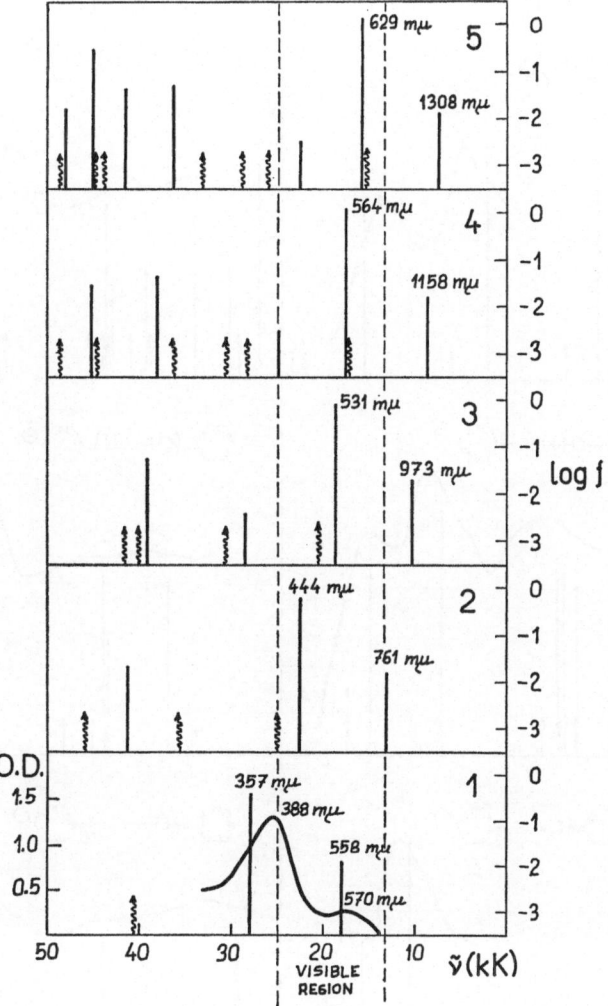

Fig. 5. Prediction of electronic spectra of radical anions of even polyenes 1—5 and the experimental absorption curve for the butadiene radical anion

cyclobutadiene), the experimental data are available for both the mono-positive and mononegative ions [62]. They are compared in Table 3 with the values inferred from the photoelectron spectral data and with the results of semiempirical calculations. Diphenylene being an alternant hydrocarbon, the PPP-like open-shell calculations give the same results for its cation and anion radicals as would be expected on the basis of pairing properties of MOs in alternant hydrocarbons [63,64]. The differences

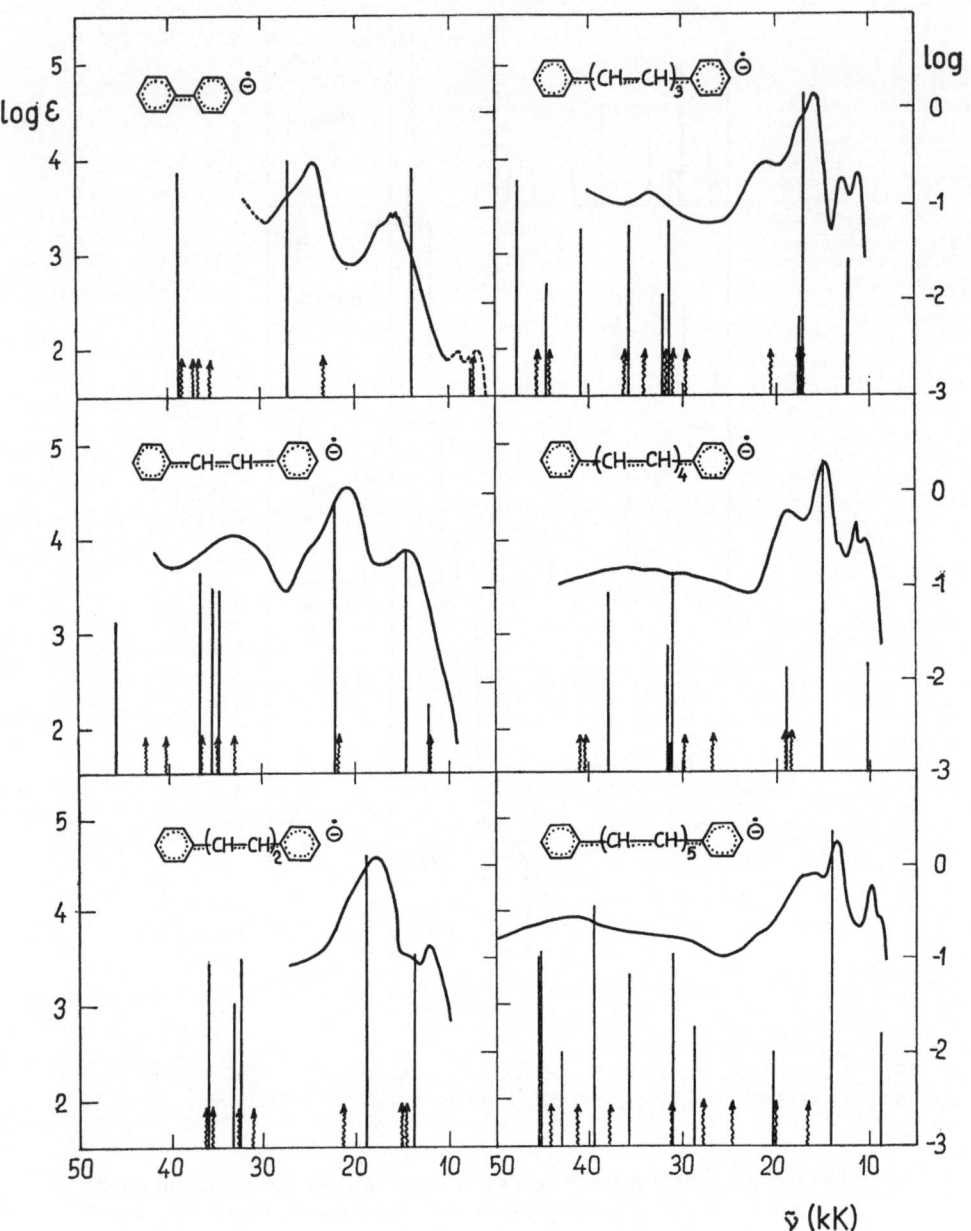

Fig. 6. Absorption curves of radical anions of α, ω-diphenylpolyenes 6—10 and results of the open-shell PPP-like calculations. Diphenyl is added for comparison

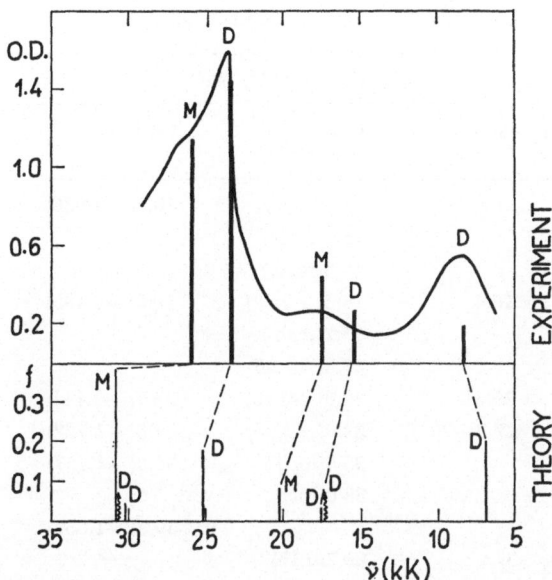

Fig. 7. Comparison of the experimental and calculated optical absorption of the monomeric (M) and dimeric (D) butadiene cation radicals. Absorption curve was recorded by Shida and Hamill, Ref. [56], the thick vertical lines represent the positions and optical density of the absorption maxima recorded by Badger and Brocklehurst, Ref. [57]. Lower part of the figure contains the results of the CNDO calculations (Čársky, P., Zahradník, R., Ref. [61])

observed in the spectra of the corresponding cation and anion radicals of alternant hydrocarbons may be assigned in part to this oversimplifying feature of the theory and in part to the solvent effect. Wasilewski [65] arrived at differing spectral characteristics for cation and anion radicals by using the orthogonalized AO basis set which brings about the breakdown of the pairing properties. The trends in his results are right, although the predicted spectral differences are somewhat overestimated.

For the 1,3,7,9,13,15,19,21-octadehydro[24]annulene radical anion (*12*; in the formula only triple bonds are indicated) two groups of bands were observed [66] in the regions at 15.6—17.8 and 21.3—25.6 kK. The semiempirical calculations [67] predict the first band in the infrared region at 2.6 kK followed by next bands at 8.6 and 10.8 kK. The electronic transition predicted at 18.5 kK can be assigned to the absorption at 15.6—17.8 kK and the next transitions at 21.6 and 24.2 kK to the absorption at 21.3—25.6 kK. The assignment is based on the assumption that the nonplanar octadehydro[24]annulene becomes planar upon the uptake of one electron, as is the case with octatetraene and its dinegative ion. The presence of the triple bonds in *12* has been

21

Table 3. Transition energies in diphenylene radical ions

Observed[1] (kK)			Calculated[2] (kK)		
PES	Cation	Anion	Cation[3] Anion[3]	Cation[4]	Anion[4]
9.4		6.6	8.9(−2.23)	10.9(−1.886)	4.8(−2.700)
			15.7(forb.)		
16.1			17.3(forb.)		
20.1	16.1	16.4	18.3(−0.82)	19.0(−0.495)	16.7(−0.456)
28.3			25.7(forb.)		23.1(−1.482)
			27.6(−1.45)		27.8(−2.200)
	27.9	25.6	30.6(−0.43)	29.1(−0.444)	28.2(−1.174)
	29.1	27.8−30.6	32.9(−2.98)	33.3(−1.722)	
			35.5(forb.)	34.0(−1.137)	36.1(−3.000)
			38.0(forb.)		
				39.3(−1.367)	
		37.2	39.7(0.15)		37.1(−0.362)

[1] Photoelectron spectral (PES) data for the monopositive ion (Eland, J. H. D., Danby, C. J., Ref. [52]) are assigned to states in which the A-type configurations dominate, the optical spectral data (Hush, N. S., Rowlands, J. R., Ref. [62]) to predicted strongly allowed transitions.
[2] Logarithms of oscillator strengths in parentheses.
[3] Open-shell PPP-like calculation according to Zahradník, R. et al., Ref. [51].
[4] After Wasilewski, J.: Dissertation, Toruń (1971).

simulated by assuming $\beta_{C \equiv C} = 1.2\, \beta_{C-C}$. Although the predicted transition energies in *12⁻* are open to some uncertainties, it is quite likely that the long-wavelength absorption has been overlooked.

The transition energies in the bismethylenecyclobutadiene radical cation (*13⁺*) have been inferred from the photoelectron spectral data and interpreted in terms of both π-electron and CNDO calculations [51].

Radical ions derived from benzenoid hydrocarbons (*14—26*) are among the best studied radicals, both experimentally and theoretically. Particularly important is the pioneering work of Hoijtink and co-workers [48,49,68] who studied theoretically and experimentally the majority of the systems considered here. Using a simple version of the CI method based on the HMO molecular orbitals and energies, they succeeded [48,68,69] in interpreting the qualitative features of the electronic spectra of radical anions. This is not surprising inasmuch as the LCAO expansion coefficients in the HMO molecular orbitals for benzenoid hydrocarbons were found [21] to be very close to those given by the SCF open-shell procedures and the HMO orbital energies were employed [70] with success

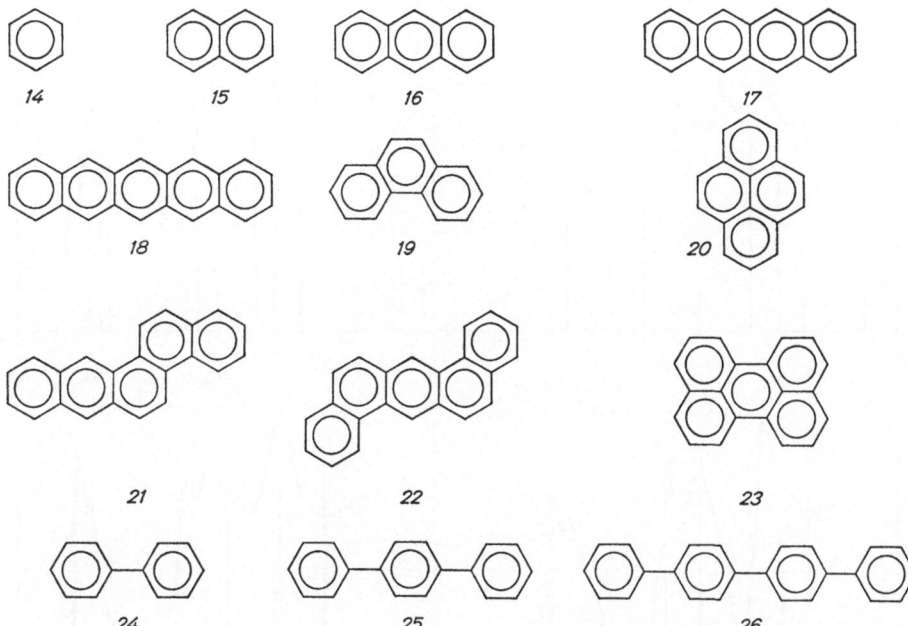

to interpret the photoelectron spectra of conjugated hydrocarbons. We even found it possible to obtain some information on the electronic spectra from the mere knowledge of HMO orbital energies and symmetries of MOs [30]. The results of semiempirical calculations as described in Section III are presented in Figs. 8 and 9. Similar calculations using somewhat different computational schemes have also been reported by other authors [37,62,65,71,72]. The semiempirical calculations also appear to give reliable predictions for the polarization directions of the absorption bands because they are in agreement with the nonempirical π-electron calculations [60] and with the available experimental data [48]. As an example, we present in Fig. 10 the results for the tetracene radical anion [30,48]. The composition of the CI wave functions indicates [30,60,65] that no meaningful interpretation of the spectra of open-shell systems is possible unless extensive allowance is made for configuration interaction, although with many systems several lowest-energy states correspond to almost pure A- and B-type electronic transitions.

The position of the benzene radical ions among the other benzenoid systems is somewhat exceptional. The presence of the degenerate frontier molecular orbitals makes it difficult to perform standard open-shell calculations because the LHP method is inherently incapable of accommodating systems of this type and the Roothaan procedure diverges here [67]. The simple CI treatments based on the HMO computational

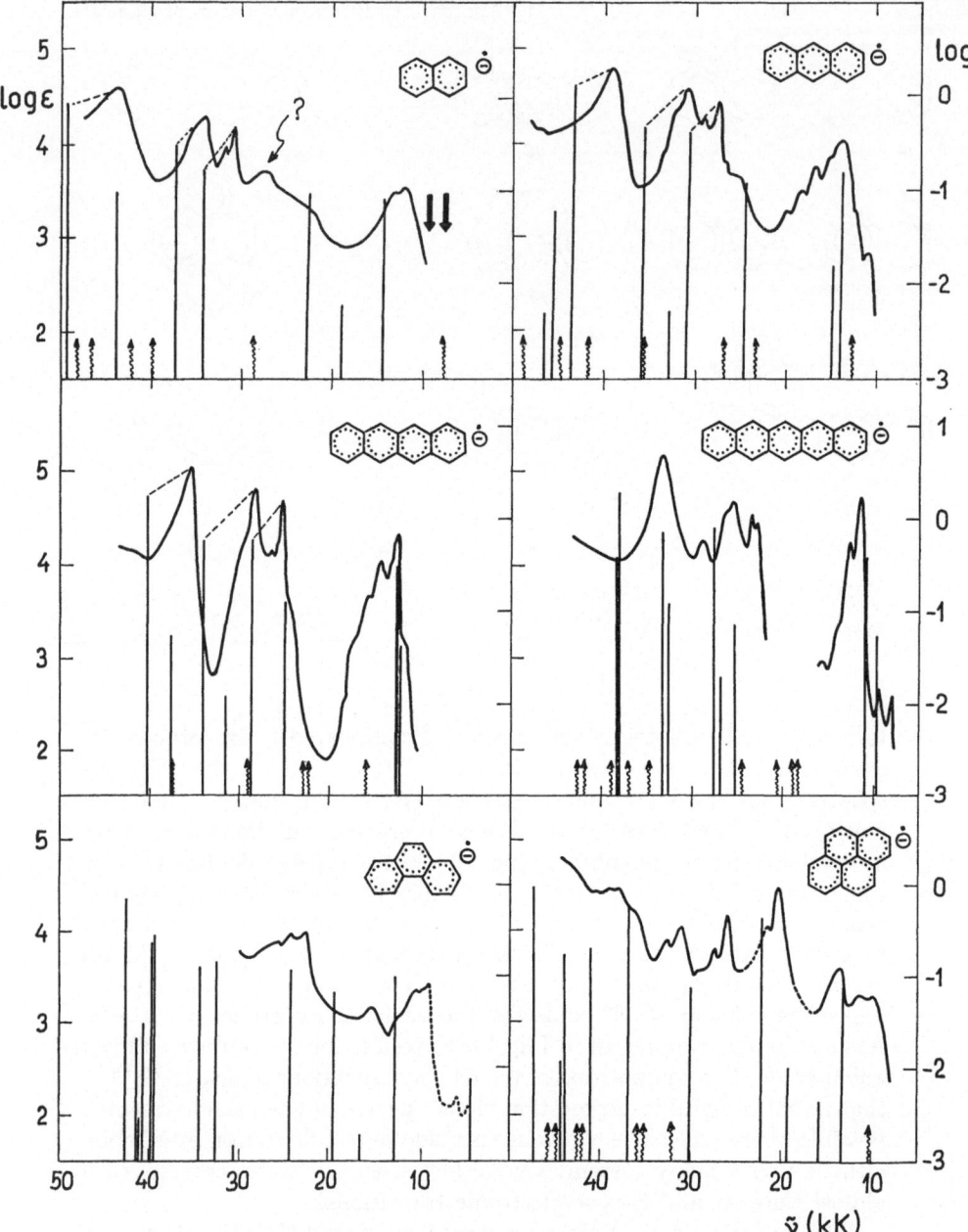

Fig. 8. Electronic spectra of the radical anions of hydrocarbons 15—20 and results of open-shell PPP-like calculations. Two arrows indicate the positions of the experimentally found long-wavelength maxima of 15⁻ (Balk, P. *et al.*, Ref. [48]). The band at 27.3 *k*K designated by a question mark was attributed to an impurity or to the protonated anion (Hoijtink, G. *et al.*, Ref. [48]). Origin of data: systems *15* and *16* (absorption curves), Brandes, K. K., Gerdes, J.: J. Phys. Chem. **71**, 508 (1967); systems *17—20* (absorption curves), Hoijtink, G. J. and coworkers, Refs. [46,48,49]; calculations, Zahradník, R., Čársky, P., Ref. [30]

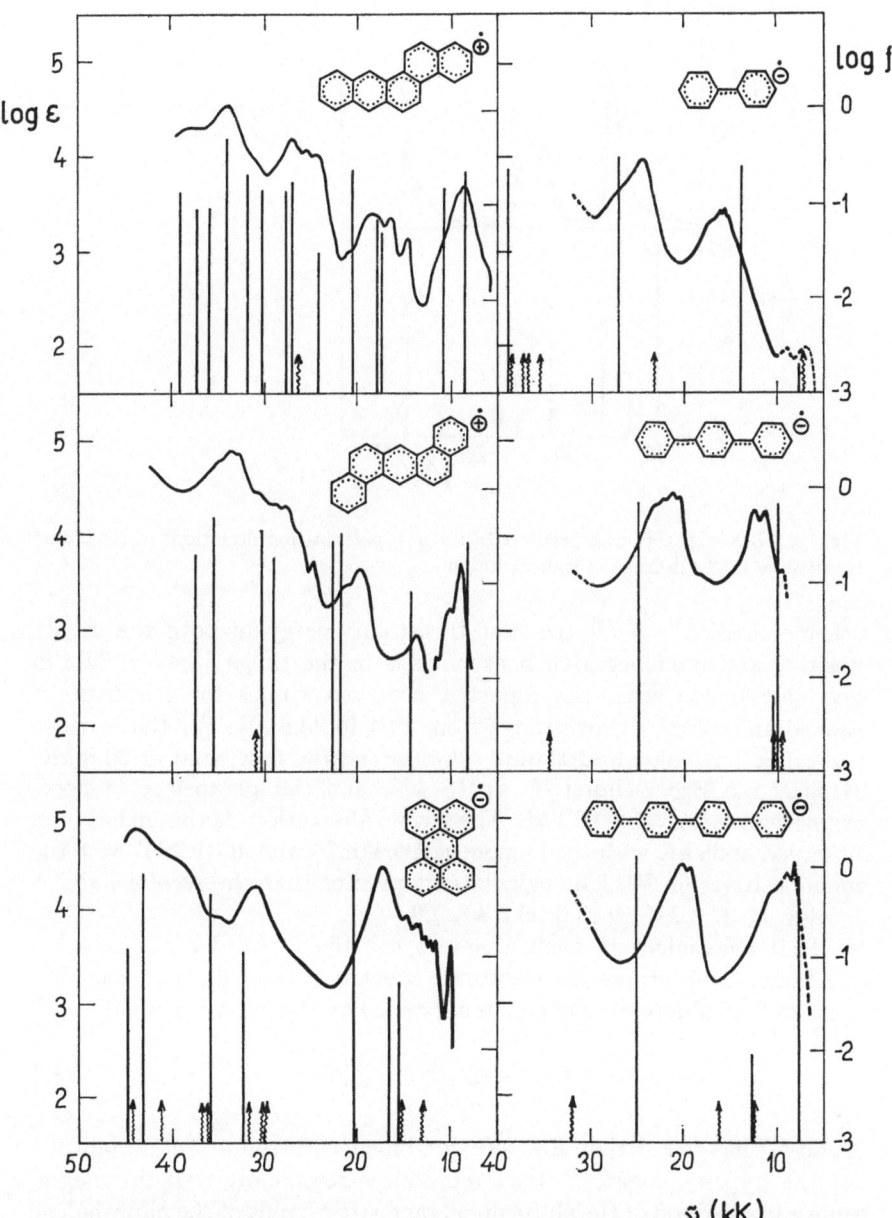

Fig. 9. Electronic spectra of the radical anions of hydrocarbons *21—26* and results of the open-shell PPP-like calculations. Origin of data: systems *21* and *22*, Distler, D., Hohlneicher, G.: Ber. Bunsenges. Physik. Chem. *74*, 960 (1970); systems *23—26* (absorption curves), Hoijtink, G. J. and coworkers, Refs. [46,48]; systems *23—26* (calculations), Čársky, P., Zahradník, R., Ref. [59]

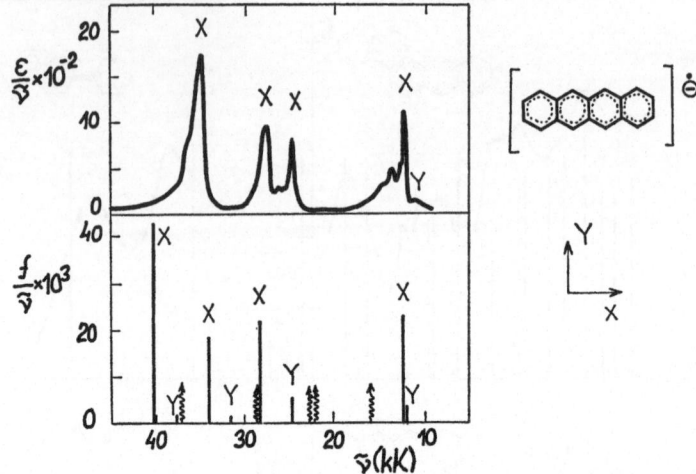

Fig. 10. Observed (top) and predicted (bottom) polarization directions of electronic transitions in the tetracene radical anion

scheme predict [69,73,74] the first transition energy in both the mono-positive and mononegative benzene ions in the range 18.4—20 kK, in good agreement with the reported observed values for the benzene radical anion [74-76] which range from 19.4 to 23.8 kK. For the benzene radical cation Shida and Hamill [77] observed the first band at 20.8 kK, Badger and Brocklehurst [78] at 18 kK, and the photoelectron spectrum [79] gives the value 18.1 kK. The second absorption maximum has been observed at 35 kK with the mononegative ion [74] and at 31.2 kK with the monopositive ion [76]. The calculations predict that the second band is located at 37.3 kK [74] and 39.5 kK [73].

It is convenient in some cases to use the first-order perturbation treatment to interpret the electronic spectra of radicals. The consider-ation of the inductive effect can be expressed by the following equation [80]

$$\Delta E_{i \to j} = \sum_{\mu} (c_{j\mu}^2 - c_{i\mu}^2) \, \delta \alpha_{\mu} \qquad (46)$$

Dodd [81] has found that Eq. (46) combined with the model accounting for the hyperconjugation effect reproduces reasonably well the magni-tude and direction of the shifts of characteristic bands of the naphthalene and anthracene radical ions upon methyl substitution. In general, with alternant hydrocarbons the shifts of the first bands upon substitution by groups with the inductive effect (e.g. CH_3 or NH_3^+) can be expected to be more considerable for radical ions than for the parent hydrocarbons, inasmuch as the first transition energy in closed-shell systems is mostly associated with the $N \rightarrow V_1$ gap, and the respective LCAO expansion coefficients in Eq. (46) are of the same absolute value. With radical ions this situation generally does not occur. The replacement of the —CH— group by nitrogen =N— can be treated in an analogous way (vide infra). Another example of the application of perturbation theory concerns the polyphenyl radical anions; here we attempted [59] to explain the increasing discrepancy between the theory and experiment for the second strong band throughout the series *24—26* (cf. Fig. 9). We assumed that this might be due to the nonplanarity of systems. Indeed, if perturbation theory is used and the bonds joining the benzene rings are treated as having less π character, a shift of the calculated transition energies towards the experimental values is found. The following expression, based on the perturbation treatment, was used:

$$\delta E_{i \rightarrow j} = 2 \sum_{\mu > \nu} \sum (c_{j\mu} c_{j\nu} - c_{i\mu} c_{i\nu}) \delta \beta_{\mu\nu} \qquad (47)$$

where all the symbols have the usual meaning.

Finally, let us mention that calculations have been reported [61,82,83] which interpreted the spectra of dimeric radical ions.

2. Alternant Odd Hydrocarbons

This group comprises several characteristic series of neutral radicals: odd polyenes, odd α,ω-diphenylpolyenes, arylmethyl radicals, and odd

$$H_2C-(CH=CH)_n-H \qquad \qquad \langle \bigcirc \rangle - \overset{\cdot}{C}H + CH=CH \overset{}{\rightarrow}_n \langle \bigcirc \rangle$$

27 (*n* = 1)	*33* (*n* = 1)
28 (*n* = 2)	*34* (*n* = 2)
29 (*n* = 3)	*35* (*n* = 3)
30 (*n* = 4)	*36* (*n* = 4)
31 (*n* = 5)	*37* (*n* = 5)
32 (*n* = 10)	

benzenoid systems of the phenalenyl type. The first member of the polyene series, the allyl radical, represents besides the benzyl radical the most popular system in theoretical open-shell studies. (These are cited in reference [31].) Here we attempt to reply to a question asked in our earlier paper [31]: "Our LCI–SCF (PPP-like) calculation results in a first strongly forbidden electronic transition at 423 mμ (this differs by only

0.8 kK from the band position reported by Currie and Ramsay [84]), and a second (allowed) transition at 197 mμ, the value of which is different by approximately 6.5 kK from the maximum given by Callear and Lee [85]. Evidently it is questionable how reasonable the π-electron approximation is in so small a molecule as allyl". The relevant answer is given in Table 4, which presents the results of the CNDO calculation

Table 4. Transition energies ($\tilde{\nu}$) and oscillator strengths (f) predicted by the CNDO calculations [67] for the allyl radical

$\tilde{\nu}(kK)$ (obsd.[84,85])	$\tilde{\nu}(kK)$	log f	Polarization direction[1]	Main configuration (type and weight)
24.5	24.6	−2.55	x	B(9−10) π°−π* 65.2
	49.5	forbidden	−	B(9−11) π°−σ* 86.1
	53.0	−1.21	z	B(9−12) π°−σ* 88.8
43.6−44.6	53.2	−0.45	x	A(8−9) π −π° 63.4

[1]) The xy plane is considered as the molecular plane.

(the description is given in Section III). It thus appears most likely that absorption in the visible can really be assigned to the π—π transition. The next three transitions fall into the range of the observed ultraviolet absorption; however, the π → π* assignment again appears to be the most likely on intensity grounds. The results of the PPP-like calculations on radicals 27—37 are presented in Fig. 11. The only experimental data available are those reported by Waterman and Dole [86] who observed the following absorption maxima for radicals 27—31 in electron beam-irradiated polyethylene (in kK): 27 (38.8); 28 (35.1); 29 (31.0); 30 (27.9); 31 (25.3). The predicted lowest-energy strongly allowed transitions are seen in Fig. 11 to reproduce these data reasonably. The transition energies given by the semiempirical full CI [29] and nonempirical [60] π-electron calculations are even in better agreement, being 1—3 kK lower than those entered in Fig. 11.

As regards the relationship between structure and colour, the neutral radicals resemble closed-shell systems rather than radical ions. This can easily be understood on the basis of the HMO orbital level schemes of these systems, in which one cannot distinguish the "unnatural" N → V gaps in radicals from the "natural" N → V_1 gaps in the corresponding closed-shell ions. Accordingly, in contrast to radical ions, absorption of neutral radicals does not extend to the extreme long-wavelength region, and the absorption maxima are shifted bathochromically upon

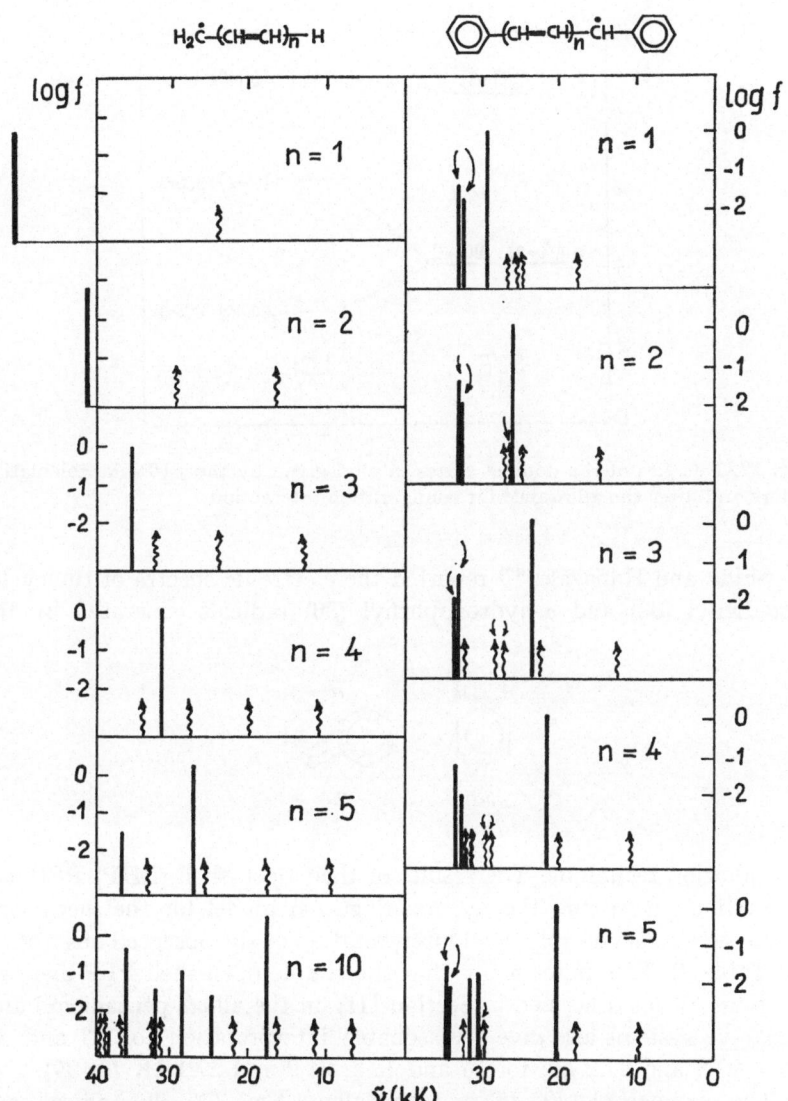

Fig. 11. Predicted electronic spectra of odd polyenes *27—32* and odd α, ω-diphenyl-polyenes *33—37*

the prolongation of the skeleton. Another characteristic feature of the odd alternant radicals is the important role of the first-order CI (cf. Fig. 12).

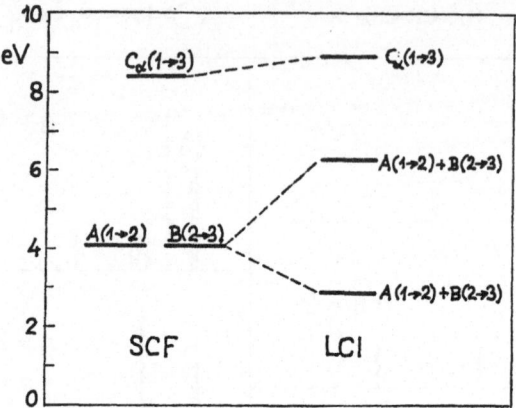

Fig. 12. Energies of the doublet states in allyl given by the PPP-like calculation before and after the allowance for configuration interaction

Shida and Hanazaki [87] reported the electronic spectra of the cyclohexadienyl (38) and α-hydronaphthyl (39) radicals generated by the

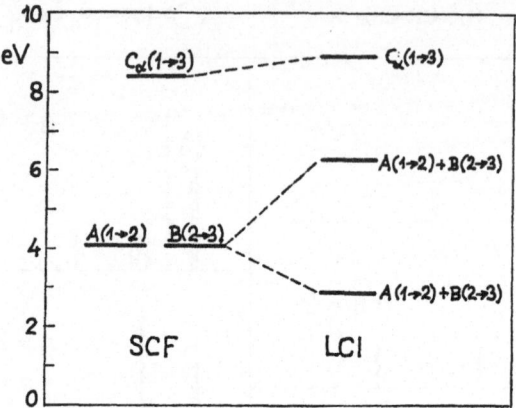

38   39

γ-irradiation technique. The results of their open-shell (LHP) PPP-like calculations, assuming the hyperconjugation model for the methylene group, give a remarkably good interpretation of the observed absorption (cf. Table 5). It is noteworthy that the simple open-shell PPP-like calculations [67] (as described in Section III) on the all-*cis* pentadienyl and cinnamyl systems also give a reasonable interpretation for 38 and 39 (16.4, 29.3 and 37.2 kK for 38 and 18.2, 24.9 and 27.1 kK for 39).

The arylmethyl (40—45) and phenalenyl-type (46—49) radicals are mentioned here only briefly. Benzyl is a radical which has been well studied both experimentally and theoretically (for a summary, see reference [31]). We have found [67] remarkable agreement between the results given by the CNDO–CI and PPP-like calculations. The CNDO–CI calculation predicts the seven lowest-energy transitions to be of the

30

Table 5. Calculated and observed spectral data for the cyclohexadienyl and α-hydronaphthyl radicals [87]

Radical	Observed $\tilde{\nu}(k\mathrm{K})$	π-Electronic SCF—LCI $\tilde{\nu}(k\mathrm{K})$	f
(cyclohexadienyl radical structure)	17.9	16.4	0.0004
		29.2	0.0003
	31.6	31.9	0.1350
		45.5	0.0420
(α-hydronaphthyl radical structure)	18.9	19.4	0.0012
	25.3	24.1	0.0003
	29.7	29.7	0.0018
		30.3	0.0538
		32.5	0.2410
		37.2	0.0018

π—π type, the next 8th transition being of the σ—π° type (at 48.8 kK, log $f = -3.5$). With the other systems 41—49 the results of π-electron calculations [31] can hardly be discussed because of the paucity of experimental data.

3. Nonalternant Hydrocarbons

Radical ions derived from nonalternant hydrocarbons have been the subject of several recent studies [88—91]. Since even such small nonalternant hydrocarbons as e.g. azulene mostly absorb in a rather long-

P. Čársky and R. Zahradník

wavelength region, the red shift of the first band on going to radical ions is usually small compared to that found with alternant hydrocarbons. It appears that with condensed systems containing five- and seven-membered rings the larger red shift is observed for the one of the two radical ions which can chemically be considered the "unnatural" form [92], *i.e.* the radical cation with molecules possessing more five-membered than seven-membered rings, and the radical anion in the opposite case.

Flash photolysis of both cyclopentadiene and ferrocene generates species with a half-life of less than 10^{-4} sec. The spectrum, consisting of two heads at 29,581 and 29,911 cm^{-1}, was assigned to the cyclopentadienyl radical [93] *(50)*. There is reasonable agreement between the observed excitation energy and the $A_2^{\cdot} \leftarrow E_1^{\cdot}$ transition at 32,500 cm^{-1} calculated semiempirically [94]. Theoretically, no other state lying below 6 eV is expected.

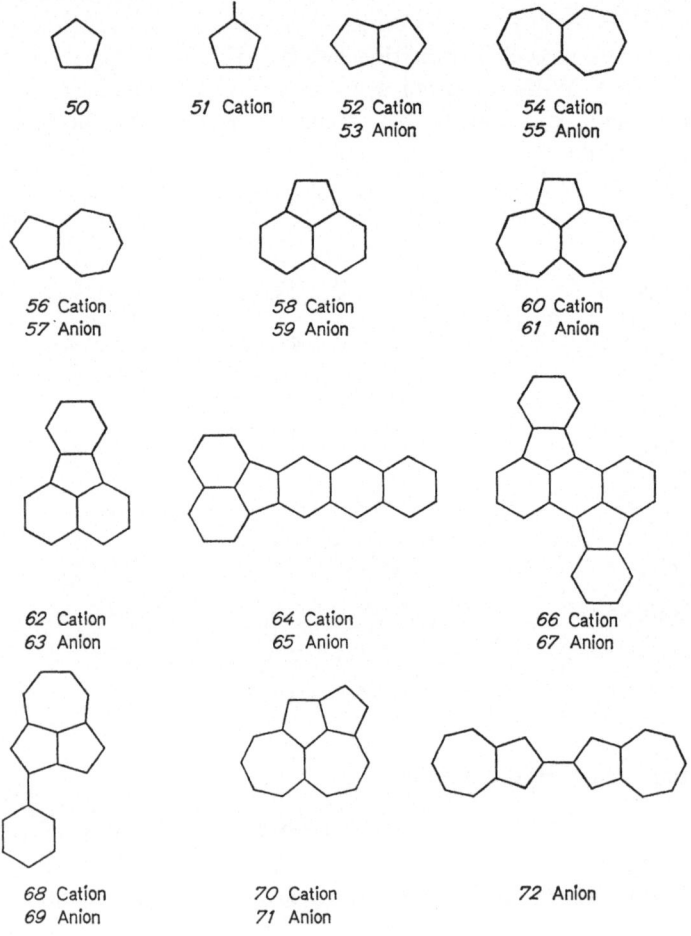

50

51 Cation

52 Cation
53 Anion

54 Cation
55 Anion

56 Cation
57 Anion

58 Cation
59 Anion

60 Cation
61 Anion

62 Cation
63 Anion

64 Cation
65 Anion

66 Cation
67 Anion

68 Cation
69 Anion

70 Cation
71 Anion

72 Anion

73 Cation   *74* Anion

The transition energies in the fulvene cation (*51*) have been inferred from the photoelectron spectra and interpreted by both π-electron and CNDO calculations [51]. In Table 6 we present some predictions for radical ions whose spectra have not yet been measured, and in Fig. 13 we present the confrontation of the π-electron calculations with experiment for a representative series of nonalternant radical ions for which absorption curves are available. The nonalternant radical ions are typical of a large number of transitions in the region 8—30 kK and accordingly of the very complex nature of electronic spectra. We also found [90] this feature with radicals *68—71*.

2,2'-Bisazulenyl radical anion (*72*) exhibits a broad, very strong band [95] (log $\varepsilon_{max} = 4.5$) in the region 14—20 kK. Its highest peak may be assigned to the allowed A-type $10 \rightarrow 11$ transition (72%). Altogether five transitions are predicted by the PPP-like calculations [67] to fall into

Table 6. First and second transition energies ($\bar{\nu}$) and oscillator strengths (f) predicted by the PPP-like calculations [67] for selected nonalternant systems

Radical	$\bar{\nu}(k\text{K})$	log f	Configurations and weights (%)
52	11.5	−3.00	A(3—4) 93
	12.9	forbidden	B(4—5) 99
53	12.9	forbidden	A(4—5) 99
	19.4	−1.00	A(3—5) 86
54	9.1	forbidden	B(6—7) 99
	15.5	−1.03	B(6—8) 87
55	7.9	forbidden	A(6—7) 99
	8.1	forbidden	B(7—8) 94
58	5.8	−2.34	A(5—6) 94
	11.2	−2.82	A(4—6) 92
60	9.9	−1.77	A(6—7) 89
	12.7	−3.23	B(7—8) 93
62	2.5	−2.57	A(7—8) 96
	9.4	−3.43	A(6—8) 97

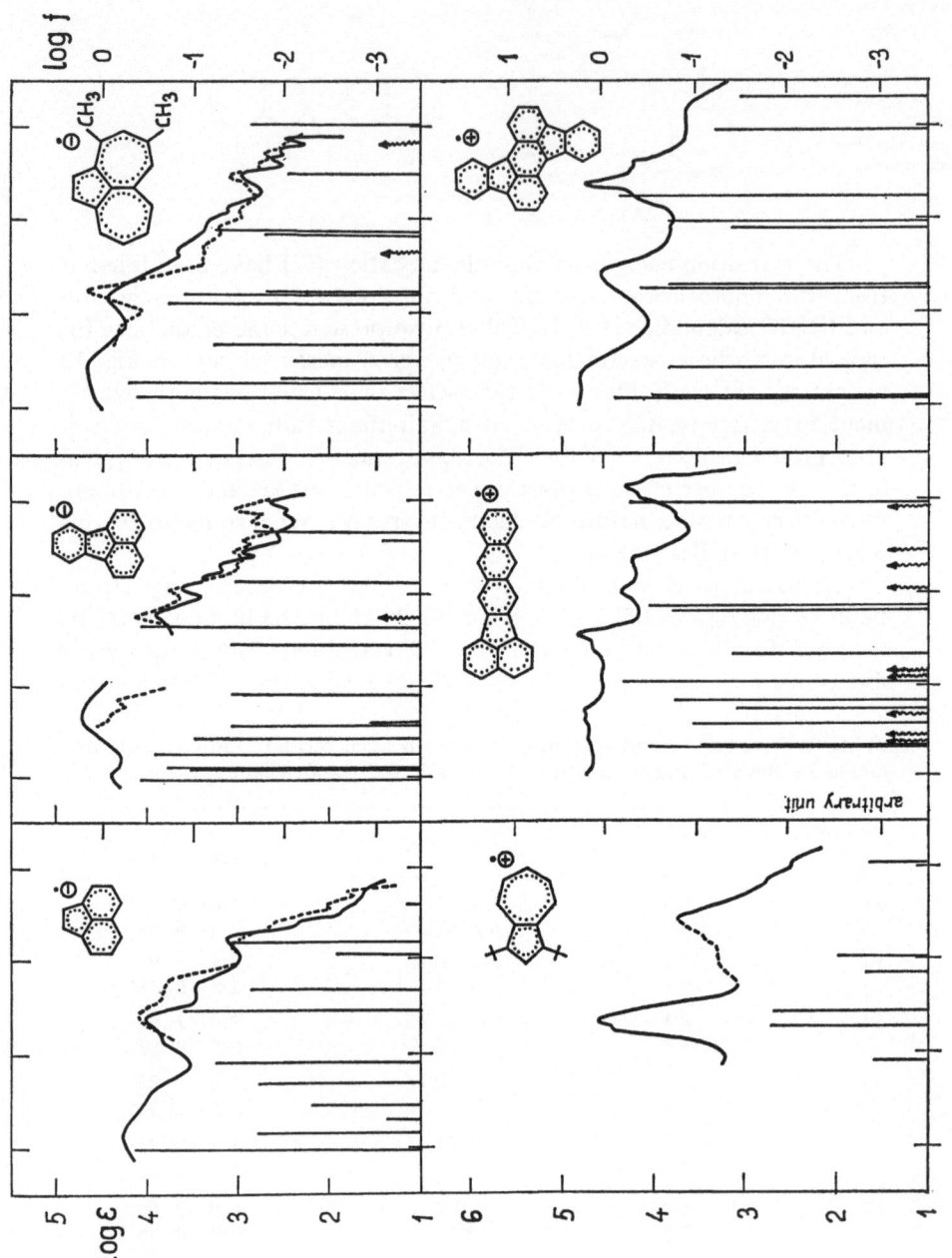

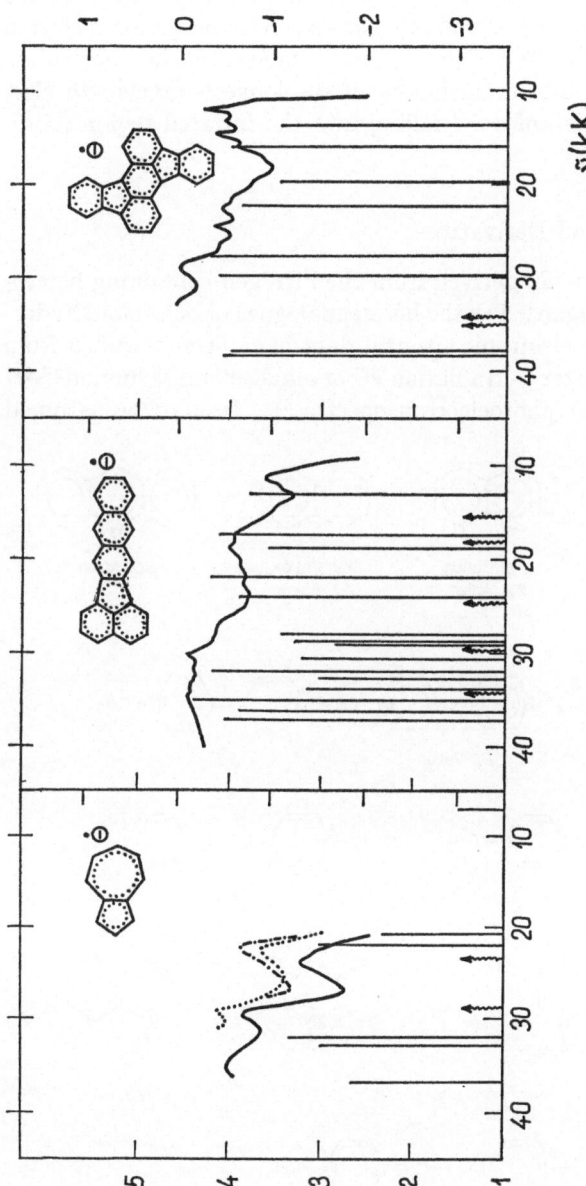

Fig. 13. Observed and predicted electronic spectra for radical ions of nonalternant hydrocarbons, Taken from: Zahradník, R. et al., Ref.[88] (systems 59, 61, 63); Nykl, I. et al., Ref.[89] (systems 56, 57); Distler, P., Hohlneicher, G., Ref.[91] (systems 64—67). In calculations the methyl groups in 61 were ignored; absorption curves drawn in differing lines refer to different techniques of radical generation (for details cf. cited papers)

the region 14—20 kK, two of them being forbidden. The theory moreover predicts two allowed transitions in the near infrared at 5.1 and 10.4 kK. The perchlorate of the cation *73* represents the first isolated salt of the hydrocarbon radical ion; π-electron calculations on its electronic spectrum were reported recently [96]. The method "molecules in molecules" afforded a reasonable interpretation [97] of the longest-wavelength electronic transition in the anion *74* falling into the infrared region (obsd. 4.9 kK, calcd. 3.6 kK).

4. Heteroanalogues and Derivatives

For the radical ions *75—82* derived from the nitrogen-containing heterocycles, which can be regarded as the heteroanalogues of benzenoid hydrocarbons, the available electronic spectral data have been obtained from direct measurements after γ-irradiation [53] or alkali-metal reduction [98,99] and inferred from the photoelectron spectra [52]. It may be assumed

75 Cation 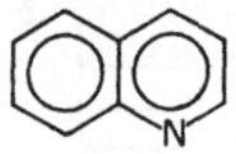   *76* Cation   *78* Cation   *80* Anion
             77 Anion   *79* Anion

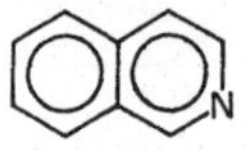

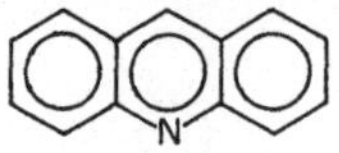

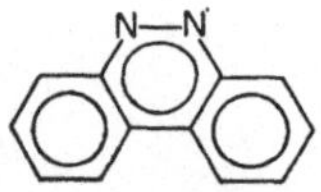

81 Anion 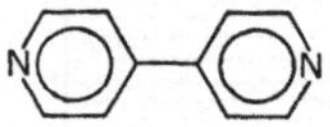   *82* Anion 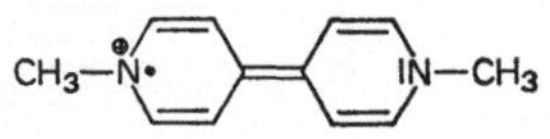   *83*

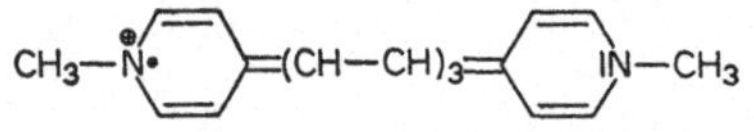

84 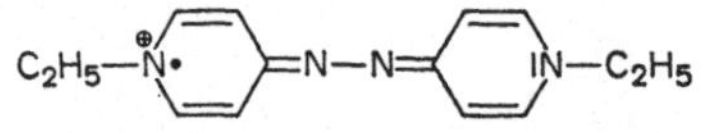   *85*

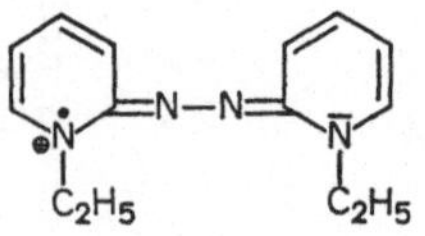

86 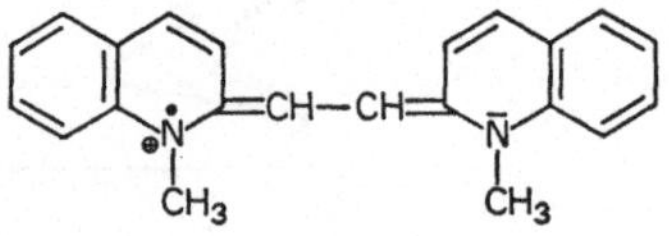   *87*

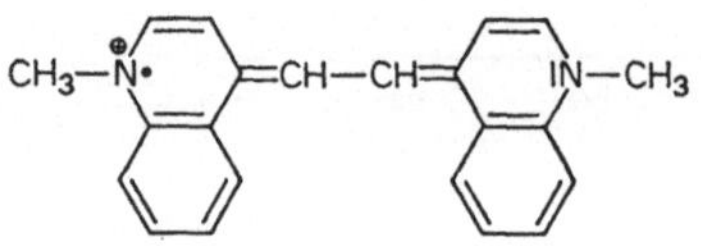

88

that the local electronic structure of nitrogen in these heterocycles is not much affected on going to the radical ions. Accordingly, we have employed in the open-shell calculations [67] the same semiempirical parameters (cf. Section III) which proved useful in standard PPP calculations on closed-shell pyridine-like heterocycles. The results are summarized in Table 7. In general, there is a similarity in the absorption spectra of radical ions of the aza compounds and the parent hydrocarbons, *e.g.* of the acridine and anthracene radical ions. With radical anions [98] the aza substitution brings about a blue shift of the first strong band, which we shall interpret here by means of first-order perturbation treatment in order to demonstrate its utility. The first strong bands of the phenanthrene and diphenyl radical anions (located [46] at 9.3 and 15.7 kK) are due to almost pure $8 \to 10$ and $7 \to 10$ transitions, respectively. Substituting [80] $\delta\alpha = -17.5$ kK into Eq. (46), one predicts for radicals

Table 7. Calculated [67] three lowest-energy electronic transitions in the radical ions *75—81* and the experimental excitation energies observed directly after γ-irradiation [53] (γ) or sodium reduction [98] (Na red.) and inferred from the photoelectron spectra [52] (PES)

Radical	Origin of data	$\bar{v}_1(k\mathrm{K})$	$\log f_1$	$\bar{v}_2(k\mathrm{K})$	$\log f_2$	$\bar{v}_3(k\mathrm{K})$	$\log f_3$
75	calcd.	5.0	−3.08	21.3	−1.63	31.8	−1.48
	γ			25			
	PES	6.1		20.3[1]		27.6[2]	
76	calcd.	6.1	−3.63	14.3	−1.10	18.6	−1.99
	γ			14.3		17.5	
	PES	4.8		16.5[3]			
77	calcd.	9.7	−3.22	15.3	−1.05	20.0	−2.62
	γ			14.0—15.2			
	Na red.			13.2—13.9			
78	calcd.	8.9	−2.71	14.3	−1.03	20.7	−2.98
	γ			14.2		21(sh.)	
	PES	7.8		15.6		24.5	
79	calcd.	7.5	−2.91	14.8	−1.06	19.2	−2.15
	γ			13.8—15.3			
	Na red.			12.8—13.9		23(sh.)	
80	calcd.	14.7	−2.18	15.7	−1.53	16.4	−0.81
	γ	12.6		14.3		17.1	
	Na red.	13.8		15.6		16.3	
81	calcd.	9.5	−1.93	17.1	−1.20	20.7	−1.39
	Na red.			14.2		22.5	

[1]) Next transition energy at 23.0 kK.
[2]) Next transition energy at 32.5 kK.
[3]) Next transition energy at 23.8 kK.

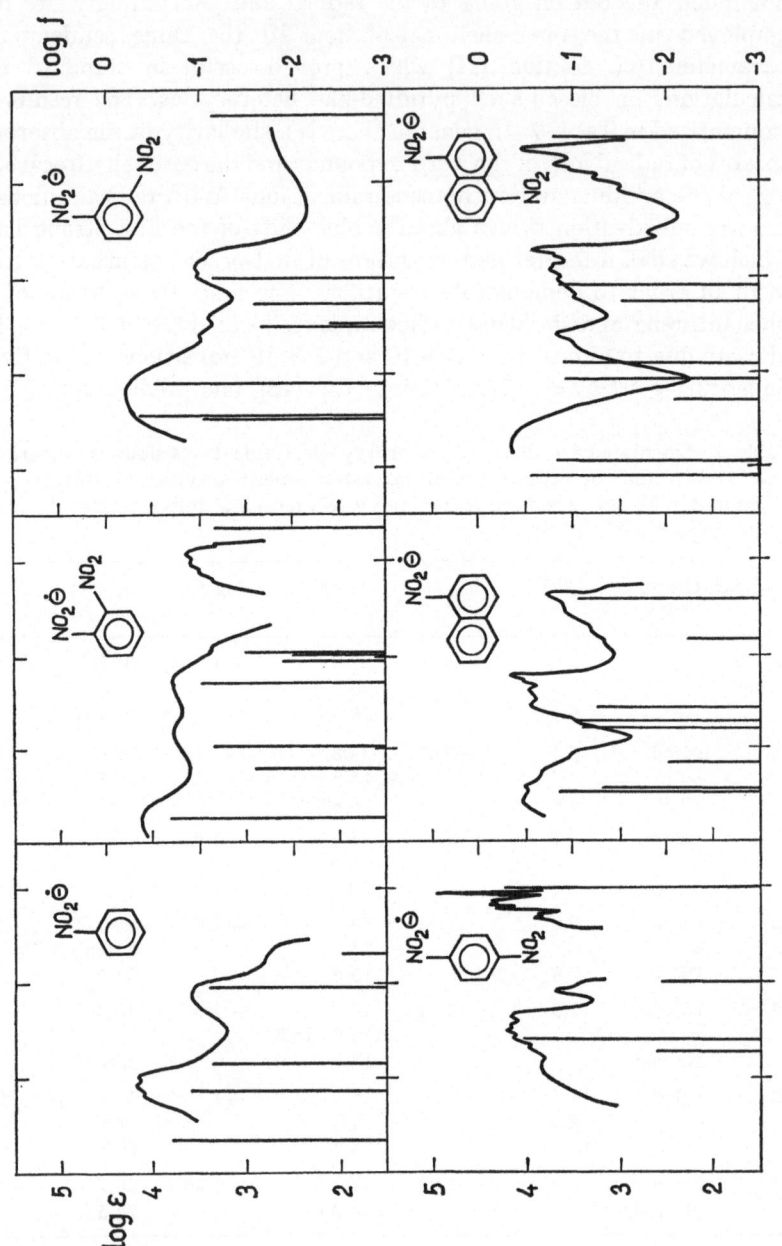

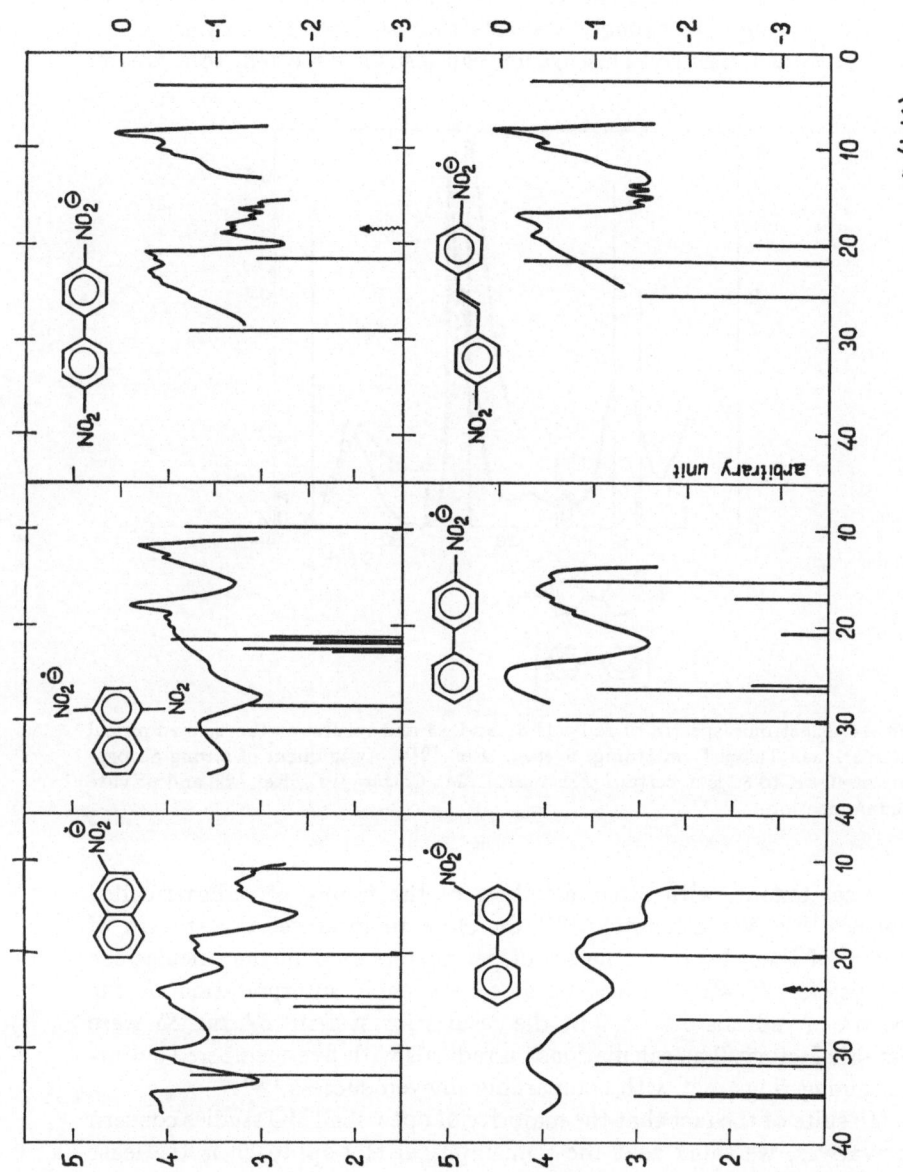

Fig. 15. Electronic spectra of aromatic nitro-substituted radical anions and results of the semiempirical calculations (taken from Shida, T., Iwata, S., Ref. 106))

81 and *82* blue shifts 4.2 and 2.9 *k*K, in reasonable agreement with the observed value [98,99] 4.9 and 1.8 *k*K.

A striking similarity in the electronic spectra [99,100] of radicals *82* and *83* (Fig. 14) strongly suggests that the two radicals are π-isoelectronic, as expected. This feature can also be expected, to a greater

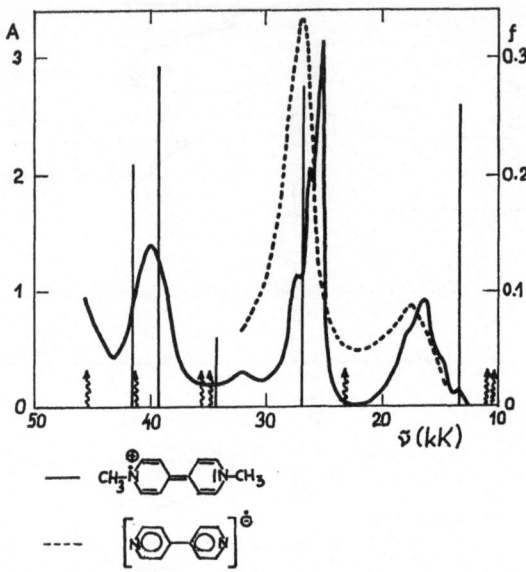

Fig. 14. Electronic spectra of radicals *82* and *83* and results of the semiempirical calculations. (Taken from Hünig, S. *et al.*, Ref. [101]). Assignment of strong absorption at 40 *k*K to *83* is uncertain (Kosower, E. M., Cotter, J. L., Ref. [100] and private communication)

or lesser extent, with other members of the family of radicals called violenes [102]. We attempted [101] therefore to interpret the spectra of several of them by the same set of parameters used in the calculations for systems *75—82*. We arrived at a reasonable interpretation of the spectra of radicals *83—86*, but the results for systems *87* and *88* were less satisfactory. The calculations on radicals with five-membered sulfur-containing rings met with comparably uneven success [103].

In spite of the fact that the majority of open-shell MO studies concern derivatives, we think that the semiempirical MO approach is the least straightforward here. In radicals such as phenoxyls, ketyls, semiquinones, or amino radicals, the odd electron is generally localized on the heteroatom, which greatly changes the local electronic structure of the latter with respect to that in the parent closed-shell molecule. Hence, in contrast to well delocalized radicals (*e.g. 75—82*), it is usually necessary to look for a new set of parameters, different from that which proved useful in treatments of the parent closed-shell molecules. This may sometimes be rather troublesome. We have encountered [104,105] such a difficulty with semiquinones, where moderate changes in parameters brought about great changes in transition energies. Hence it may happen that the set of parameters fitted to the electronic spectrum of a single radical will fail to reproduce the spectra of other radicals of the same structural type. Any semiempirical study should therefore cover several radicals of a given type. A very good example of such a treament is a recent study of Shida and Iwata [106], who recorded the spectra of a large series of radical anions derived from nitro derivatives of benzene, naphthalene, diphenyl, and stilbene, and interpreted them by the type of calculations described in Section III, using the following parameters [107]: $I_N = 30.38$ eV, $I_O = 19.25$; all the β's were made proportional to the overlap integrals. The results for the twelve selected nitro radicals are presented in Fig. 15. Generally, however, the theoretical studies have been devoted to one particular radical (cf. Table 8), and sometimes the parameters were adjusted according to the observed transition energies.

C. CNDO Calculations on Small Radicals

The number of semiempirical all-valence electron studies devoted to electronic spectra of radicals is still limited. The LHP–CNDO/2 treatment combined with configuration interaction was used by Kikuchi [108] to interpret the electronic spectrum of H_2NO. King *et al.* [2] calculated the four lowest doublet states of the SO_3F radical by the Pople-Nesbet [109] unrestricted procedure using the standard CNDO computational scheme; the results are consistent with the observed spectrum. We have attempted a systematic study on small radicals for which spectral data were available. On the basis of experience accumulated with closed-shell systems [110,111] we selected the computational scheme of Del Bene and Jaffé [38]. Besides the calculations of radicals presented in Table 9, we have also made [22] predictions for HNCN and BOF_2 radicals. In spite of the approximate nature of these calculations, the results appear to be useful, making it possible not only to analyze and classify the electronic transitions but also to provide further useful information. For example, with BH_2 the theory affords an explanation as to why no band has been observed in the near-ultraviolet region; with HO_2, the calculations predict a weak band in the infrared or long-wavelength visible region. Nevertheless the calculations have encountered some difficulties, as is only to be expected with any semiempirical treatment.

Table 8. π-Electronic calculations on electronic spectra of derivatives of conjugated hydrocarbons

Radical	Description of the calculation	Transition energies[1] (eV) and oscillator strengths (in parentheses)
Anilino	Open-shell Roothaan SCF + CI (singly excited states); Pariser method for two center γ integrals; parameters (in eV): $I_N - I_C = 1.677$; $\beta_{CC} = -2.395$; $\beta_{CN} = -2.576$, $\gamma_{CC} = 11.400$, $\gamma_{NN} = 12.799$	obsd.: 3.12 ; 4.04 calcd.[2]: 2.83(0.0065); 3.50(0.0171); 4.27(0.4705); 4.72(0.5291)
Phenoxyl	As with the anilino radical; $I_O - I_C = 3.449$, $\beta_{CO} = -3.000$, $\gamma_{OO} = 14.657$	obsd.: 3.10 ; 4.23 calcd.[2]: 2.46(0.0003); 3.43(0.0428); 4.16(0.6265); 4.71(0.2163)
Nitrosobenzene anion	LHP SCF + CI (singly excited states); Pariser-Parr approximation for γ integrals; parameters (in eV): $I_C = 11.22$; $I_N = 14.51$; $I_O = 17.795$; $\beta_{CC} = -2.39$, $\beta_{CN} = -1.89$, $\beta_{NO} = -1.27$, $\gamma_{CC} = 10.60$, $\gamma_{NN} = 13.31$, $\gamma_{OO} = 14.65$	obsd.: 2.032 ; 2.980 calcd.[3]: 2.647(0.0426); 2.742(0.0602); 2.952(0.0814)
Nitrobenzene anion	As with the nitrosobenzene anion, $I_N = 28.855$, $I_O = 17.764$, $\beta_{CN} = -2.40$, $\beta_{NO} = -2.40$, $\gamma_{NN} = 16.595$, $\gamma_{OO} = 13.894$	obsd.: 2.214 ; 4.052 calcd.[3]: 2.321(0.0063); 2.685(0.119); 4.486(0.0011); 4.731(0.0255)
Benzonitrile anion	As with the nitrosobenzene anion, parameters for the cyano group (in eV): $I_C = 11.24$, $I_N = 14.47$, $\beta_{CC'} = -2.802$, $\beta_{CN} = -3.280$, $\gamma_{CC} = 10.66$, $\gamma_{NN} = 13.26$	obsd.: 1.653 ; 2.530 ; 3.263 calcd.[3]: 0.713(0.002); 2.228(0.2231); 3.472(0.0086)

Phthalonitrile anion	As with the benzonitrile anion	obsd.[3]:	1.240	2.138	2.362	3.646
		calcd.[3]:	0.619(0.0148);	2.461(0.1234);	2.839(0.0454);	4.392(0.0672z)
Isophthalonitrile anion	As with the benzonitrile anion	obsd.[3]:	(1.653)?	2.175	3.542	4.959
		calcd.[3]:	0.350(0.0373);	1.974(0.1753);	3.175(0.0504)[4];	4.661(0.1069)
Terephthalonitrile anion	As with the benzonitrile anion	obsd.[3]:	2.480	3.646		
		calcd.[3]:	1.881(0.3866);	4.020 and 4.345(0.2196)		
Pyromellitonitrile anion	As with the benzonitrile anion	obsd.[3]:	2.684	3.306 and 3.492 4.444	4.779	
		calcd.[3]:	1.941(0.3366);	3.379(0.0635);	4.413(0.3012);	4.592(0.0127)
p-Phenylenediamine cation	LHP SCF + CI (singly excited states), parameters (in eV): $I_C = 11.42$, $\beta_{CC} = \beta_{CN} = -2.39$, $I_N = 23.39$ (adjusted)	obsd.[5]:	2.50(0.11);	3.34(0.06);	3.80(0.31);	5.08(0.1)
		calcd.[5]:	2.64(0.63)[6];	3.45(0.18);	3.66(0.22);	4.76(0.06)
Tetramethyl-p-phenylenediamine cation	Roothaan open-shell SCF, parameters for N adjusted: $I_C = 11.16$ eV, $I_N = 26.04$ eV, $\beta_{CC} = -2.39$ eV, $\beta_{CN} = -2.25$ eV	obsd.:	3.90	4.73		
		calcd.[7]:	4.173(0.296);	5.030(0.956)		
Tetracyanoquinodimethane anion	Roothaan open-shell SCF + CI (5 singly excited states)	obsd.[8]:	1.49[8]	1.66[8]	2.94[8]	4.40[9] 5.33[9]
		calcd.[8]:	all 7 parameter sets give one transition at ~1.5 eV, the second one at ~3.1 eV			
	Roothaan open-shell SCF, 3 parameter sets	calcd.[9]:	good agreement with experiment except for transition at 3.62 eV (0.156) not identified experimentally			

Table 8 (continued)

Radical	Description of the calculation	Transition energies[1] (eV) and oscillator strengths (in parentheses)
	As with the nitrosobenzene anion, $I_O = 34.4$ eV, $\beta_{CO} = -3.42$ eV, $\gamma_{OO} = 19.46$ eV	obsd.: 2.99 4.59 calcd.[10]: 2.835(0.005); 2.988(0.022); 4.313(0.038)[11]; 4.900(0.22)
	As with the preceding radical	obsd.: 2.28(0.065) 3.76(0.32) calcd.[10]: 2.580(0.080)[12]; 4.048(0.022); 4.121(0.040); 4.192(0.42)
Perfluoro-2,1,3-benzoselenodiazole anion	Roothaan open-shell SCF + CI	obsd.: 2.17 and 2.85(0.04); 3.45(0.13); 4.46(0.06) calcd.[13]: 1.90(0.008); 2.83(0.076); 3.39(0.221); 4.10(0.081) obsd.: 5.27(0.30) calcd.[13]: 4.40(0.541); 5.52(0.042)
Diphenylpicryl-hydrazyl	Roothaan open-shell SCF + CI (5 singly excited states)	obsd.: 2.39 3.81 calcd.[14]: 2.43(0.501)[15]; 3.29(0.161); 3.53(0.205); 4.00(0.170)

Footnotes to Table 8

---

[1] For references to experimental data, see the papers cited.
[2] Hinchliffe, A., Stainbank, R. E., Ali, M. A.: Theoret. Chim. Acta 5, 95 (1966); predicted intensities are squares of transition moments.
[3] Ishitani, A., Nagakura, S.: Ref. [37].
[4] Next calculated transition at 4.177(0.0402).
[5] Kimura, K., Mataga, N.: J. Chem. Phys. 51, 4167 (1969).
[6] The first transition predicted at 1.10 eV is forbidden.
[7] Monkhorst, H. J., Kommandeur, J.: J. Chem. Phys. 47, 391 (1967).
[8] Lowitz, D. A.: J. Chem. Phys. 46, 4698 (1967).
[9] Jonkman, H. T., Kommandeur, J.: Chem. Phys. Letters 15, 496 (1972).
[10] Kanamaru, N., Nagakura, S.: Bull. Chem. Soc. Japan 43, 3443 (1970).
[11] Next transition predicted at 4.717(0.020).
[12] Next three weak ($f < 0.005$) transitions are predicted 2.952, 2.959, and 3.502 eV.
[13] Fajer, M. et. al.: Ref. [27].
[14] Gubanov, V. A., Pereliaeva, L. A., Chirkov, A. K., Matevosian, R. O.: Theoret. Chim. Acta 18, 177 (1970).
[15] Next weak transition is predicted at 2.58(0.072).

We have found [40] a drastic failure with N_2^+. While with CO^+ the interpretation of the spectrum was successful, with the isoelectronic N_2^+ radical the theory not only underestimates the first transition energy, it also predicts that the $^2\Pi$ state will be of lower energy than the $^2\Sigma_g^+$ (SCF ground) state. Incidentally, the incorrect order of the two states is also predicted by ab initio calculations [112,113].

D. Ab Initio Calculations

As this topic alone could be the subject of an extensive review, no great effort was devoted to making the coverage of the literature complete, to discussing the specific problems encountered in ab initio calculations, or to analyzing the results attained. We present rather in Table 10 a representative set of results of ab initio calculations. The object here is to demonstrate the present state in this field and at the same time to facilitate comparison with semiempirical all-valence electron calculations.

VI. Concluding Remarks

In spite of the rapid development in computer facilities, it appears that the π-electron approximation will remain for a long time to come the only manageable approach for the interpretation of the electronic spectra of large conjugated radicals. For this reason we think that attempts to

45

P. Čársky and R. Zahradník

Table 9. CNDO calculations on the electronic spectra of small radicals [22,40]

Radical		Wave numbers (kK) and oscillator strengths					
		$\tilde{\nu}_1$	$\log f_1$	$\tilde{\nu}_2$	$\log f_2$	$\tilde{\nu}_3$	$\log f_3$
BeH	calcd.	16.19	−1.19	60.58	−1.17	−70.50	−1.22
	obsd.	20.04					
BH^+	calcd.	21.98	−1.46	70.55	−1.22		
	obsd.	26.38					
BO	calcd.	34.74	−1.80	39.98	−1.29	50.95	−1.53
	obsd.	23.90				43.17	
CN	calcd.	1.48	−3.30	32.27	−1.21	38.14	−2.42
	obsd.	9.11		25.75			
CO^+	calcd.	16.58	−2.25	44.86	−1.25	49.95	−1.67
	obsd.	20.41		45.63			
BH_2	calcd.	7.96	−2.24	61.22	−1.94	72.76	−1.83
	obsd.	11.6−15.6					
NH_2	calcd.	22.60	−2.09	65.09	forbidden	79.13	−1.59
	obsd.	11.1−23.3					
HCO	calcd.	15.08	−2.42	35.98	−2.70	36.70	−1.49
	obsd.	11.6−21.7		24.4−38.5			
HNF	calcd.	25.24	−2.11	73.39	−2.16	75.86	−1.50
	obsd.	20.00−25.64					
HO_2	calcd.	3.12	forbidden	62.48	−0.88	64.71	−2.00
	obsd.			34.5−51.3			
NO_2	calcd.	11.41	−2.11	12.08	forbidden	13.00	−3.21
	obsd.			10.0−31.2			

		$\tilde{\nu}_4$	$\log f_4$	$\tilde{\nu}_5$	$\log f_5$	$\tilde{\nu}_6$	$\log f_6$
	calcd.	23.16	forbidden	30.51	forbidden	55.14	forbidden
NF_2	calcd.	38.48	−1.98	63.81	forbidden	69.97	forbidden
	obsd.	35.7−42.5					
FO_2	calcd.	9.90	forbidden	47.05	−2.77	66.08	−1.88
	obsd.	16.7−18.2	ε<100	>33			
FCO	calcd.	15.05	−2.40	35.75	−2.60	−37.10	−1.56
	obsd.	18.55				29.4−45.5	
H_2NO	calcd.	10.80	forbidden	50.36	−1.79		
	obsd.[1]	22.00−24.50	0.9[2]	~43.50	~3.5[2]		
F_2CN	calcd.	17.14	forbidden	26.35	forbidden	26.58	forbidden

		$\tilde{\nu}_4$	$\log f_4$	$\tilde{\nu}_5$	$\log f_5$	$\tilde{\nu}_6$	$\log f_6$
	calcd.	36.15	−2.02	52.03	−1.65	52.36	−2.11
	obsd.	27.6−29.5					

[1] Experiment for $(t\text{-}C_4H_9)_2NO$.
[2] $\log \varepsilon$.

46

Table 10. Ab initio calculations on electronic spectra of radicals

Radical	Computational method	Basis set[1]		Electronic transitions[2]
CH	SCF method of equivalence and symmetry restrictions[3] applied to ground and several excited states	Large basis developed[4] for CH($^2\Pi$)	obsd.[5]:	2.88 3.18 3.94
			calcd.[5]:	2.94($A^2\Delta \leftarrow X^2\Pi$); 3.25($B^2\Sigma^- \leftarrow X^2\Pi$); 3.97($C^2\Sigma^+ \leftarrow X^2\Pi$); 7.29 7.31 ($D^2\Pi \leftarrow X^2\Pi$)
NH$^+$	As with CH	Large basis developed[4] for NH($^3\Sigma^-$)	obsd.[5]:	0.04 2.67 2.81
			calcd.[5]:	0.06($a^4\Sigma^- \leftarrow X^2\Pi$); 2.78($A^2\Sigma^- \leftarrow X^2\Pi$); 2.90($B^2\Delta \leftarrow X^2\Pi$); 4.26 4.28 ($C^2\Sigma^+ \leftarrow X^2\Pi$)
CO$^+$	Limited CI treatment (several configurations) using virtual orbitals; different SCF Hamiltonians for σ and π MOs	Double-zeta basis with 3d orbitals[6]	obsd.[7]:	
			calcd.[7]:	0.30($^2\Pi(1) \leftarrow A^2\Pi$); 4.68($^2\Pi(2) \leftarrow A^2\Pi$); 5.24 7.47($^2\Delta \leftarrow A^2\Pi$)
N$_2^+$	Roothaan SCF method applied to $^2\Sigma_g^+$, $^2\Pi_u$ and $^2\Sigma_u^+$ states	As with CO$^+$ Symmetry STF basis	obsd.[7]:	5.41
			calcd.[7]:	4.25($D^2\Pi_g \leftarrow A^2\Pi_u$); 7.46($^2\Delta_u(1) \leftarrow A^2\Pi_u$); 7.83($^2\Delta_u(2) \leftarrow A^2\Pi_u$)
			calcd.[8]:	incorrect order of the $^2\Pi_u$ and $^2\Sigma_g^+$ (ground) states
CN	As with CO$^+$	As with CO$^+$	obsd.[7]:	5.61 6.27 6.94
			calcd.[7]:	8.30($D^2\Pi \leftarrow A^2\Pi$); 7.66($F^2\Delta \leftarrow A^2\Pi$); 9.05($J^2\Delta \leftarrow A^2\Pi$)

47

Table 10 (continued)

Radical	Computational method	Basis set[1]	Electronic transitions[2]
CN	Full valence CI calculation	Hartree-Fock AOs for the inner-shell electrons of C and N, STF with optimized exponents for 2s and 2p AOs	obsd.: 1.146 3.193 6.755 calcd.[9]: 1.883($A^2\Pi$II ← $X^2\Sigma^+$I); 3.765($B^2\Sigma^+$II ← $X^2\Sigma^+$I); ~6.593($D^2\Pi$III ← $X^2\Sigma^+$I); obsd.: 7.556 7.334 calcd.[9]: 7.742($^2\Phi$I ← $X^2\Sigma^+$I); 7.807($H^2\Pi$IIII ← $X^2\Sigma^+$I); 7.856($E^2\Sigma^+$III ← $X^2\Sigma^+$I); obsd.: 7.400 8.090 calcd.[9]: 7.879($F^2\Delta$I ← $X^2\Sigma^+$I); 7.988($^2\Sigma^-$I ← $X^2\Sigma^+$I); 8.432($J^2\Delta$II ← $X^2\Sigma^+$I); obsd.: 8.704($^2\Sigma^-$II ← $X^2\Sigma^+$I); ~8.76($^2\Pi$IV ← $X^2\Sigma^+$I)
CF	As with CH	3s, p, and d functions included	obsd.: 5.32 6.12 calcd.[10]: 5.04($A^2\Sigma^+$ ← $X^2\Pi$); 6.32($B^2\Delta$ ← $X^2\Pi$)
BH_2	Natural orbital CI (singly and doubly excited states)	Contracted GTF basis comparable to "double zeta" STF set	obsd.: 0.53 calcd.[11]: 0.28($^2B_1$ ← $^2A_1$)
NH_2	1, SCF; 2, multi-configuration first-order wave functions treatment using the iterative natural-orbital procedure	Contracted [4s 2p 1d/ 2s 1p] GTF basis	obsd.: ~1.4 (adiabatic) calcd.[12]: $^2A_1$ ← $^2B_1$, difference in energy minima, 1.41(SCF), 1.67 (CI)

NH$_2$	SCF OCBSE method[13] applied to ground and several excited states	STO-3G[14] 4-31G[16]	obsd.: 1.38–2.88 calcd.[15]: 2.92(^{2}A$_1$ ← ^{2}B$_1$) calcd.[15]: 1.78(^{2}A$_1$ ← ^{2}B$_1$)
HCO	Limited CI based on the SCF MOs given by the method of Nesbet[3]	Combination of several lobe GTF bases published	obsd.: 1.44–2.69 calcd.[17]: 2.45(^{2}A′ ← ^{2}A″)
NF$_2$	SCF OCBSE method[13] applied to ground and several excited states	STO-3G[14]	obsd.: 4.43–5.28 calcd.[15]: 4.24(^{2}A$_1$ ← ^{2}B$_1$)
NO$_2$	As with NF$_2$ Single-configuration SCF, "indirect" method of McWeeny[18] As with CH	STO-3G[14] (5s 2p) GTF basis[19] (7s 3p) GTF basis[19] Lobe GTF expansions of near-Hartree-Fock AOs[21]	obsd.: 1.24–3.87; 4.80–5.27 calcd.[15]: 1.68(^{2}A$_2$ ← ^{2}A$_1$); 3.15(^{2}B$_2$ ← ^{2}A$_1$) calcd.[20]: 2.27(^{4}B$_2$ ← ^{2}A$_1$); 2.33(^{4}A$_2$ ← ^{2}A$_1$); 2.43(^{2}B$_1$ ← ^{2}A$_1$); 3.09(^{2}A$_2$ ← ^{2}A$_1$); 4.20(^{2}B$_2$ ← ^{2}A$_1$) calcd.[20]: 2.32(^{4}B$_2$ ← ^{2}A$_1$); 2.29(^{4}A$_2$ ← ^{2}A$_1$); 2.50(^{2}B$_1$ ← ^{2}A$_1$); 2.98(^{2}A$_2$ ← ^{2}A$_1$); 3.97(^{2}B$_2$ ← ^{2}A$_1$) calcd.[22]: 2.13(^{2}B$_1$ ← ^{2}A$_1$); 3.95(^{2}B$_2$ ← ^{2}A$_1$)

49

Table 10 (continued)

Radical	Computational method	Basis set[1]	Electronic transitions[2]
CH_3	SCF method of Roothaan applied to ground and excited states	(9 s 5p/4 s) GTF set[23] plus additional terms to represent the $n=2$ shell of H and $n=3$ shell of C	obsd.: 5.73 calcd.[24]: 5.40(3s Rydberg $^2A_1' \leftarrow {}^2A_2''$); 6.59($3p_{xy}$ Rydberg $^2E' \leftarrow {}^2A_2''$); 7.60(valence $^2E' \leftarrow {}^2A_2''$)
Allyl	Large CI calculations (119 $^2A_1$, 111 $^2A_2$, 106 $^2B_1$, and 120 $^2B_2$ states) based on MOs of $C_3H_5^+$	Lobe GTF expansions of near-Hartree-Fock AOs[21]	obsd.: experimental data range from 3.04 to 5.0 eV calcd.[25]: 3.79($^2B_1 \leftarrow {}^2A_2$); 6.49($^4A_2 \leftarrow {}^2A_2$); 7.91($^2B_2 \leftarrow {}^2A_2$); 8.06($2^2B_1 \leftarrow {}^2A_2$)
SO_3F	EWSD SCF method[26]; virtual orbital approximation	STO-3G[14] but different exponents for F, sulfur 3d AOs included	obsd.: ~0.91; 1.61; 2.40 calcd.[26]: 2.50($^2E \leftarrow {}^2A_2$); 2.90($^2E \leftarrow {}^2A_2$); 3.66($^2A_1 \leftarrow {}^2A_2$); 5.19($^2A_1 \leftarrow {}^2A_2$)

Footnotes to Table 10

[1] STF and GTF stand for Slater-type function and Gaussian-type function bases.
[2] In eV; for references to experimental data see theoretical papers cited.
[3] Nesbet, R. K.: Proc. Roy. Soc. (London) A 230, 312 (1955).
[4] Cade, P. E., Huo, W. M.: J. Chem. Phys. 47, 614 (1967).
[5] Liu, H. P. D., Verhaegen, G.: J. Chem. Phys. 53, 735 (1970).
[6] Nesbet, R. K.: J. Chem. Phys. 40, 3619 (1964).
[7] Guérin, F.: Theoret. Chim. Acta 17, 97 (1970).
[8] Cade, P. E., Sales, K. D., Wahl, A. C.: J. Chem. Phys. 44, 1973 (1966).
[9] Schaefer III, H. F., Heil, T. G.: J. Chem. Phys. 54, 2573 (1971); only doublet-doublet transitions are tabulated.
[10] Hall, J. A., Richards, W. G.: Mol. Phys. 23, 331 (1972).
[11] Bender, C. F., Schaefer III, H. F.: J. Mol. Spectry. 37, 423 (1971).
[12] Bender, C. F., Schaefer III, H. F.: J. Chem. Phys. 55, 4798 (1971).
[13] Hunt, W. J., Dunning, T. H., Jr., Goddard III, W. A.: Chem. Phys. Letters 3, 606 (1969).
[14] Hehre, W. J., Stewart, R. F., Pople, J. A.: J. Chem. Phys. 51, 2657 (1969).
[15] Del Bene, J. E.: J. Chem. Phys. 54, 3487 (1971).
[16] Ditchfield, R., Hehre, W. J., Pople, J. A.: J. Chem. Phys. 54, 724 (1971).
[17] Fink, W. H.: J. Am. Chem. Soc. 95, 1073 (1972).
[18] Berthier, G.: In: Molecular orbitals in chemistry, physics and biology (eds. P. Löwdin and B. Pullman). New York: Academic Press 1964.
[19] Orbital exponent result from an optimization carried out on the isolated atoms by C. J. Hornbach (Ph. D. thesis, Cleveland 1967).
[20] Burnelle, L., May, A. M., Gangi, R. A.: J. Chem. Phys. 49, 561 (1968); only the five lowest-energy transitions are tabulated.
[21] Whitten, J. L.: J. Chem. Phys. 44, 359 (1966).
[22] Fink, W. H.: J. Chem. Phys. 49, 5054 (1968).
[23] Huzinaga, S.: J. Chem. Phys. 42, 1293 (1965).
[24] McDiarmid, R.: Theoret. Chim. Acta 20, 282 (1971).
[25] Peyerimhoff, S. D., Buenker, R. J.: J. Chem. Phys. 51, 2528 (1969).
[26] Hillier, I. H., Saunders, V. R.: Intern. J. Quantum Chem. 4, 503 (1970).

further refine the open-shell π-electron theory are still worthwhile. The situation with semiempirical all-valence electron methods is rather unsatisfactory because there is no method capable of giving reasonable transition energies and at the same time reliable potential surfaces[a]. The standard methods (e.g. CNDO/2, INDO) that give good energy predictions for ground states fail to reproduce the electronic spectra, whereas the CNDO method of Del Bene and Jaffé that gives good transition energies cannot be used to predict geometry or for similar problems. With respect to the feasibility of calculations, the further development of the semiempirical all-valence electron theory appears to be highly

[a] This has been found for closed-shell systems [111] but it very likely applies to radicals, too.

desirable. Ab initio calculations give results closely matching the experimental values, provided that large GTF, and in some cases also CI basis sets, are used; this, however, makes the theoretical approach economically prohibitive. This explains why the results of highest accuracy have been obtained with diatomic radicals.

Notes added in proof

1. Dr. Gey (Berlin) found our classification of open shell SCF methodes somewhat inaccurate in distinguishing between the methodes using coupling operators and the density matrix method inasmuch as the latter only represents a more elegant formulation of the former. This is certainly true, but from the practical point of view the two computational techniques are different so that our classification is perhaps still reasonable.

2. We wish to mention further papers relevant to the theory of SCF methods using coupling operators: Dyadyuscha, G. G., Kuprievich, V. A.: Teor. Eksp. Khim. *1*, 406 (1967). — Kuprievich, V. A.: Intern. J. Quant. Chem. *1*, 561 (1967). — Kruglyak, Yu. A., Dyadyuscha, G. G.: Theoret. Chim. Acta *12*, 18 (1968). — Kruglyak, Yu. A., Mozdor, E. V., Kuprievich, V. A.: Croat. Chem. Acta *43*, 15 (1971).

3. In a recent paper (Albat, R., Gruen, N.: Chem. Phys. Letters *18*, 572 (1973)) the OCBSE and related techniques have been demonstrated not to satisfy all necessary conditions for the energy to be stationary: this shortcoming manifests itself as a dependence of the self-consistent result on the initial input guess molecular orbitals (see also Basch, H., McKoy, V.: J. Chem. Phys. *53*, 1628 (1970) and Macaulay, R., Goutier, D.: Chem. Phys. Letters *18*, 501 (1973)). The open shell SCF procedure reported recently (Peters, D.: J. Chem. Phys. *57*, 4351 (1972)) is actually the OCBSE method.

4. In the survey of π-electron semiempirical studies we have missed a paper of Karwowski, J. (Bull. Acad. Polon. Sci., Ser. Sci. Math. Astronom. Phys. *20*, 413 (1972)) on the interpretation of the electronic spectra of benzene ion radicals. Recently Shida, T. and Iwata, S. (J. Am. Chem. Soc. *95*, 3473 (1973)) recorded spectra of a large series of ion radicals of aromatic hydrocarbons and analyzed them in terms of π-electron SCF CI calcu!ations. They confirmed experimentally a long-wavelength transition in the mono-negative azulene ion predicted by the theory. Their absorption curve of the monopositive azulene ion is in better agreement with the calculation than the curve presented in Fig. 31; a strong absorption maximum at 15 kK in the latter is thus likely due to other species than the azulene radical cation. From recent ab initio studies we wish mention those on CH (Lie, G. C., Hinze, J., Liu, B.: J. Chem. Phys. *57*, 625 (1972) and Walker, T. E. H., Kelly, H. P.: J. Chem. Phys. *57*, 936 (1972)), BH_2 (Staemmler, V., Jungen, M.: Chem. Phys. Letters *16*, 187 (1972)), HO_2 (Gole, J. L., Hayes, E. F.: J. Chem. Phys. *57*, 360 (1972)), and on CO_2^- (Krauss, M., Neumann, D.: Chem. Phys. Letters *14*, 27 (1972)).

5. Recently we have extended the SCF-CI computational scheme (described in Chapter 3) for accomodation of radicals having a two-fold degenerate MO occupied by one or three electrons (Kuhn, J., Čársky, P., Zahradník, R.: Theoret. Chim. Acta, to be submitted for publication).

We wish to thank Drs. Gey and Sauer (Berlin) for reading the manuscript and valuable comments and Mrs. Žohová and Mrs. Týleová for a technical assistance in preparation of the manuscript.

VII. References

1) Berthier, G.: In: Molecular orbitals in chemistry, physics and biology (eds. P. Löwdin and B. Pullman). New York: Academic Press 1964.
2) King, G. W., Santry, D. P., Warren, C. H.: J. Chem. Phys. *50*, 4565 (1969).
3) Roothaan, C. C. J.: Rev. Mod. Phys. *32*, 179 (1960).
4) Kroto, H. W., Santry, D. P.: J. Chem. Phys. *47*, 2736 (1967).
5) Chang, S. Y., Davidson, E. R., Vincow, G.: J. Chem. Phys. *52*, 1740 (1970).
6) Adams, O. W., Lykos, P. G.: J. Chem. Phys. *34*, 1444 (1961).
7) Zahradník, R., Čársky, P.: Progr. Phys. Org. Chem., in the press.
8) Huzinaga, S.: Phys. Rev. *120*, 866 (1960).
9) Huzinaga, S.: Phys. Rev. *122*, 131 (1961).
10) Birss, F. W., Fraga, S.: J. Chem. Phys. *38*, 2553 (1963).
11) Sleeman, D. H.: Theoret. Chim. Acta *11*, 135 (1968).
12) Claxton, T. A., Smith, N. A.: Theoret. Chim. Acta *22*, 399 (1971).
13) Čársky, P., Zahradník, R.: Theoret. Chim. Acta *26*, 171 (1972).
14) Hunt, W. J., Dunning, T. H., Jr., Goddard III, W. A.: Chem. Phys. Letters *3*, 606 (1969).
15) Segal, G. A.: J. Chem. Phys. *53*, 360 (1970).
16) Chang, S. Y., Davidson, E. R., Vincow, G.: J. Chem. Phys. *52*, 5596 (1970).
17) McWeeny, R.: Proc. Roy. Soc. (London) *A241*, 239 (1957).
18) Hillier, I. H., Saunders, V. R.: Intern. J. Quantum Chem. *4*, 503 (1970).
19) Nesbet, R. K.: Proc. Roy. Soc. (London) *A230*, 312 (1955).
20) Longuet-Higgins, H. C., Pople, J. A.: Proc. Phys. Soc. (London) *A68*, 591 (1955).
21) Čársky, P., Zahradník, R.: Collection Czech. Chem. Commun. *36*, 961 (1971).
22) Zahradník, R., Čársky, P.: Theoret. Chim. Acta *27*, 121 (1972).
23) Bender, C. F., Davidson, E. R.: J. Phys. Chem. *70*, 2675 (1966).
24) Bender, C. F., Davidson, E. R.: J. Chem. Phys. *46*, 3313 (1967).
25) Bender, C. F., Schaefer III, H. F.: J. Chem. Phys. *55*, 4798 (1971).
26) Bender, C. F., Schaefer III, H. F.: J. Mol. Spectry. *37*, 423 (1971).
27) Fajer, J., Bielski, B. H. J., Felton, R. H.: J. Phys. Chem. *72*, 1281 (1968).
28) Čársky, P., Zahradník, R.: Theoret. Chim. Acta *17*, 316 (1970).
29) Amos, T., Woodward, M.: J. Chem. Phys. *50*, 119 (1969).
30) Zahradník, R., Čársky, P.: J. Phys. Chem. *74*, 1240 (1970).
31) Čársky, P., Zahradník, R.: J. Phys. Chem. *74*, 1249 (1970).
32) Brabant, C., Salahub, D. R.: Theoret. Chim. Acta *23*, 285 (1971).
33) Peyerimhoff, S. D., Buenker, R. J.: J. Chem. Phys. *51*, 2528 (1969).
34) Dewar, M. J. S., Hashmall, J. A., Venier, C. G.: J. Am. Chem. Soc. *90*, 1953 (1968).
35) Dewar, M. J. S., Trinajstić, N.: Chem. Commun. *1970*, 646.
36) Ellison, F. O., Matheu, F. M.: Chem. Phys. Letters *10*, 322 (1971).
37) Ishitani, A., Nagakura, S.: Theoret. Chim. Acta *4*, 236 (1966).
38) Del Bene, J., Jaffé, H. H.: J. Chem. Phys. *48*, 1807 (1968).
39) Ellis, R. L., Kuehnlenz, G., Jaffé, H. H.: Theoret. Chim. Acta *26*, 131 (1972).
40) Čársky, P., Macháček, M., Zahradník, R.: Collection Czech. Chem. Commun., submitted for publication.
41) Habersbergerová, A., Janovský, I., Teplý, J.: Radiation Res. Rev. *1*, 109 (1968); Habersbergerová, A., Janovský, I., Kouřím, P.: Radiation Res. Rev. *4*, 123 (1972).
42) Land, E. J.: In: Progress in reaction kinetics (ed. G. Porter), Vol. 3. Oxford: Pergamon Press 1965.

43) Hamill, W. H.: In: Radical ions (eds. E. T. Kaiser and L. Kevan). New York: Interscience Publishers 1968.
44) Herzberg, G.: Molecular spectra and molecular structure. I. Spectra of diatomic molecules. New York: Van Nostrand 1959.
45) Herzberg, G.: Molecular spectra and molecular structure. III. Electronic spectra and electronic structure of polyatomic molecules. New York: Nostrand Reinhold Comp. 1966.
46) Balk, P., Hoijtink, G. J., Schreurs, J. W. H.: Rec. Trav. Chim. 76, 813 (1957).
47) Hoijtink, G. J., van der Meij, P. H.: Z. Physik. Chem. (Frankfurt) 20, 1 (1959).
48) Hoijtink, G. J., Velthorst, N. H., Zandstra, P. J.: Mol. Phys. 3, 533 (1960).
49) Buschow, K. H. J., Dieleman, J., Hoijtink, G. J.: J. Chem. Phys. 42, 1993 (1965).
50) UV atlas of organic compounds. Weinheim: Verlag Chemie; London: Butterworths 1966.
51) Zahradník, R., Čársky, P., Slanina, Z.: Collection Czech. Chem. Commun. 38, 1886 (1973).
52) Eland, J. H. D., Danby, C. J.: Z. Naturforsch. 23a, 355 (1968).
53) David, C., Janssen, P., Geuskens, G.: Spectrochim. Acta 27A, 367 (1971).
54) Brugman, C. J. M., Rettschnick, R. P. H., Hoytink, G. J.: Chem. Phys. Letters 8, 263 (1971).
55) Zahradník, R., Michl, J.: Collection Czech. Chem. Commun. 30, 515 (1965).
56) Shida, T., Hamill, W. H.: J. Am. Chem. Soc. 88, 5371 (1966).
57) Badger, B., Brocklehurst, B.: Trans. Faraday Soc. 65, 2576 (1969).
58) Shida, T., Hamill, W. H.: J. Chem. Phys. 44, 4372 (1966).
59) Čársky, P., Zahradník, R.: Collection Czech. Chem. Commun. 35, 892 (1970).
60) Roberts, G., Warren, K. D.: Theoret. Chim. Acta 22, 184 (1971).
61) Čársky, P., Zahradník, R.: Theoret. Chim. Acta 20, 343 (1971).
62) Hush, N. S., Rowlands, J. R.: Mol. Phys. 6, 317 (1963).
63) Coulson, C. A., Rushbrooke, G. S.: Proc. Cambridge Phil. Soc. 36, 193 (1940).
64) McLachlan, A. D.: Mol. Phys. 2, 271 (1959).
65) Wasilewski, J.: Dissertation, Toruń (1971).
66) McQuilkin, R. M., Garratt, P. J., Sondheimer, F.: J. Am. Chem. Soc. 92, 6682 (1970).
67) Čársky, P., Zahradník, R.: Calculations for this review.
68) Balk, P., DeBruijn, S., Hoijtink, G. J.: Rec. Trav. Chim. 76, 907 (1957).
69) Hoijtink, G. J.: Mol. Phys. 2, 85 (1959).
70) Brogli, F., Heilbronner, E.: Theoret. Chim. Acta 26, 289 (1972).
71) Hinchliffe, A., Murrell, J. N., Trinajstić, N.: Trans. Faraday Soc. 62, 1362 (1966).
72) Wasilewski, J.: Acta Phys. Polonica 38A, 349 (1970).
73) Ishitani, A., Kuwata, K., Tsubomura, H., Nagakura, S.: Bull. Chem. Soc. Japan 36, 1357 (1963).
74) Gardner, C. L.: J. Chem. Phys. 45, 572 (1966).
75) Ishitani, A., Nagakura, S.: Mol. Phys. 12, 1 (1967).
76) Ekstrom, A.: J. Phys. Chem. 74, 1705 (1970).
77) Shida, T., Hamill, W. H.: J. Chem. Phys. 44, 2375 (1966).
78) Badger, B., Brocklehurst, B.: Trans. Faraday Soc. 65, 2582 (1969).
79) Turner, D. W.: Advan. Phys. Org. Chem. 4, 31 (1966).
80) Murrell, J. N.: The theory of the electronic spectra of organic molecules, p. 252. London: Methuen 1963.
81) Dodd, J. W.: J. Chem. Soc. (B) Phys. Org. 1971, 2427.
82) Badger, B., Brocklehurst, B.: Trans. Faraday Soc. 66, 2939 (1970).

83) Shida, T., Iwata, S., J. Chem. Phys. 56, 2858 (1972).
84) Currie, C. L., Ramsay, D. A.: J. Chem. Phys. 45, 488 (1966).
85) Callear, A. B., Lee, H. K.: Trans. Faraday Soc. 64, 308 (1968).
86) Waterman, D. C., Dole, M.: J. Phys. Chem. 74, 1906 (1970).
87) Shida, T., Hanazaki, I.: Bull. Chem. Soc. Japan 43, 646 (1970).
88) Zahradník, R., Rejholec, V., Hobza, P., Čársky, P., Hafner, K.: Collection Czech. Chem. Commun. 37, 1983 (1972).
89) Nykl, I., Fojtík, A., Hobza, P., Čársky, P., Zahradník, R., Shida, T.: Collection Czech. Chem. Commun. 38, 1459 (1973).
90) Nykl, I., Rejholec, V., Hobza, P., Čársky, P., Zahradník, R., Hafner, K.: Collection Czech. Chem. Commun. 38, 1463 (1973).
91) Distler, D., Hohlneicher, G.: Ber. Bunsenges. Physik. Chem. 74, 960 (1970).
92) Zahradník, R., Párkányi, C., Michl, J., Horák, V.: Tetrahedron 22, 1341 (1966).
93) Thrush, B. A.: Nature 178, 155 (1956).
94) Longuet-Higgins, H. C., McEwen, K. L.: J. Chem. Phys. 26, 719 (1957).
95) Ikegami, Y., Seto, S.: Bull. Chem. Soc. Japan 43, 2409 (1970).
96) Čársky, P., Chalvet, O., Hünig, S., Scheutzow, D., Zahradník, R.: Collection Czech. Chem. Commun. 36, 560 (1971).
97) Brugman, C. J. M., van Asselt, N. P., Rettschnick, R. P. H., Hoytink, G. J.: Theoret. Chim. Acta 23, 105 (1971).
98) Chaudhuri, J., Kume, S., Jagur-Grodzinski, J., Szwarc, M.: J. Am. Chem. Soc. 90, 6421 (1968).
99) Schmulbach, C. D., Hinckley, C. C., Wasmund, D.: J. Am. Chem. Soc. 90, 6600 (1968).
100) Kosower, E. M., Cotter, J. L.: J. Am. Chem. Soc. 86, 5524 (1964).
101) Hünig, S., Scheutzow, D., Čársky, P., Zahradník, R.: J. Phys. Chem. 75, 335 (1971).
102) Hünig, S.: Pure Appl. Chem. 15, 109 (1967).
103) Zahradník, R., Čársky, P., Hünig, S., Kiesslich, G., Scheutzow, D.: Intern. J. Sulfur Chem. C6, 109 (1971).
104) Čársky, P., Hobza, P., Zahradník, R.: Collection Czech. Chem. Commun. 36, 1291 (1971).
105) Hobza, P., Čársky, P., Zahradník, R.: Collection Czech. Chem. Commun. 38, 641 (1973).
106) Shida, T., Iwata, S.: J. Phys. Chem. 75, 2591 (1971).
107) Shida, T.: Private communication (1972).
108) Kikuchi, O.: Bull. Chem. Soc. Japan 42, 47 (1969).
109) Pople, J. A., Nesbet, R. K.: J. Chem. Phys. 22, 571 (1954).
110) Jaffé, H. H.: Accounts Chem. Res. 2, 136 (1969).
111) Klopman, G., O'Leary, B.: Fortschr. Chem. Forsch. 15, 445 (1970).
112) Cade, P. E., Sales, K. D., Wahl, A. C.: J. Chem. Phys. 44, 1973 (1966).
113) Rose, J. B., McKoy, V.: J. Chem. Phys. 55, 5435 (1971).

Received February 26, 1973

Ion Cyclotron Resonance Spectroscopy

Professor Dr. Hermann Hartmann, Dr. Karl-Heinz Lebert and Dr. Karl-Peter Wanczek

Institut für Physikalische Chemie der Universität Frankfurt am Main

Contents

Introduction

Ion cyclotron resonance (ICR) is the excitation of the cyclotron frequency of ion motion in a magnetic field by an electric rf field at resonance. ICR spectrometry is based on the application of the resonance phenomenon to the detection of ions.

Ion cyclotron resonance spectroscopy is responsible for some of the increase in publications about ion–molecule collision processes during the last decade. Although ion cyclotron resonance has been known in principle for some time, specially designed apparatus for scientific applications has only recently become available. The technique is now well established and the ion cyclotron double-resonance (ICDR) technique, in particular, has furnished the means for rapid surveys of interesting ion–molecule interactions. The flexibility of ICR techniques makes them suitable for many applications.

The existing reviews [36,116,147,161,162] all deal with the operation of ICR and its applications to problems of chemistry and physics.

It is the purpose of this article to explain the basic principle of ion cyclotron resonance, to describe in detail its practical realization in an ICR spectrometer, and to explain the different techniques that have been developed. Because ICR spectrometry differs in some important respects from conventional mass spectrometry, some remarks about the information conveyed by ICR signals are added. The attempt is made to summarize previously reported results both in descriptions and in tables of data on ion–molecule reactions (IMR) as observed with ICR techniques. There is a very complete bibliography of publications dealing with ICR, although it was not possible to review all of the papers listed there.

I. Basic Principles and Instrumentation

A. Ion Motion

Ions of mass m and charge q move in a uniform magnetic field B in circular orbits with the cyclotron frequency

$$\omega_c = qB/m. \tag{1}$$

The motion is constrained by the magnetic field in a plane perpendicular to B but is unconstrained parallel to B.

If a static electric field E is superimposed normal to the field B, a drift motion is superimposed on the circular motion of the ions. In this crossed-field geometry the ions move in cycloidal orbits with the cyclotron frequency and drift with the velocity

$$v_d = E/B \qquad (2)$$

in the direction perpendicular to $\vec{B}$ and $\vec{E}$. This can be derived from the equation of motion

$$\dot{\vec{v}} = \frac{q}{m} \left[\vec{E} + \vec{v} \times \vec{B} \right], \qquad (3)$$

which has the solution

$$\begin{cases} x = \dfrac{E}{B} t + A \sin \omega_c t - \dfrac{v_y}{\omega_c} \left(1 - \cos \omega_c t\right) \\[2ex] y = \dfrac{v_y}{\omega_c} \sin \omega_c t + A \left(1 - \cos \omega_c t\right) \end{cases} \qquad (4)$$

where v_x and v_y are the components of initial velocity and $A = (v_x/\omega_c - E/B\omega_c)$. Eq. (4) describes a cycloidal path.

This periodic motion can be used to establish a swarm of ions drifting through a cell, for example, of the three-sectioned standard type which is shown schematically in Fig. 1. In the source region, ions are generated from a gas by impact with an electron beam which crosses the cell parallel to the magnetic field. The top and bottom plates of the cell establish the electric drift field normal to a magnetic field in the source

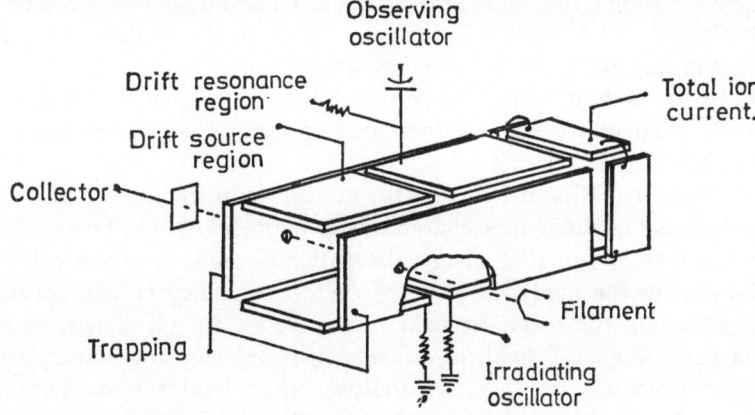

Fig. 1. Cutaway view of an ICR cell (flat, three-section cell)

and analyzer region and the side plates constitute a trapping field, so that ion motion is essentially restricted to the center of the cell. Finally, a third region, free of electric fields, serves as collector for the ions. Both positive and negative ions produced in the source pass through the cell; their separation is achieved by appropriate potentials to the trapping plates.

B. Ion Spectra

The essential feature of ion cyclotron resonance is the resonance excited by the periodic motion of the ions as they drift through the cell. If an rf electric field $E_1(t)$ of frequency ω_1 is applied between the drift plates of the analyzer region, ions can absorb energy from the rf field provided the resonance condition $\omega_1 = \omega_c$ is fulfilled. As a consequence of energy absorption, the ions are continuously accelerated to larger orbital radii until the process is terminated by collision with either another particle or a wall. The cyclotron resonance frequency ω_c and the rf field frequency ω_1 can be used for mass analysis of the ion swarm. B and E are held constant, and the total ion current (TIC) is monitored while the frequency of a strong rf electric field is swept, so that ions of specific cyclotron frequency ω_c — and hence of specific mass — are excited and collide with the drift plates of the cell at resonance. This sweep-out of ions is monitored as a reduction of the TIC. Thus one obtains a spectrum of reductions of TIC; the relative reductions are proportional to the number of ions of given mass in the ion swarm passing the cell.

The cyclotron motion of ions can also be excited by a weak field without sweeping them out from the swarm, which enables the absorbed energies at the different resonance frequencies to be obtained. This can be done by using either an rf bridge circuit whose imbalance in resonance is displayed, or a marginal oscillator, extremely sensitive to small changes in impedance in the resonant circuit.

Both methods of obtaining ion cyclotron resonance conditions can be used, particularly for plotting mass spectra. They are respectively termed TIC or ICR spectra.

While both TIC and ICR spectra are linear in the mass scale, plots of ion intensities differ in a characteristic manner. As will be explained later, the intensity of ICR energy absorption depends on the transit time of the ions in the analyzer. If the detector frequency is held constant while sweeping the magnetic field intensity, as Eq. (2) shows, ions of higher mass (higher B field) have lower drift velocity and consequently a longer residence time in the analyzer than lighter ions. The ICR spectrum hence shows higher peak intensities in the high mass range. To compare ion intensities I by means of power absorptions A, a mass correction must be performed. Since currents of ions of different mass are independent of either ω_c or B, the reduction R in the TIC is a direct measure of the contribution of a given ion to the total ion current, obviating mass correction. Signals of primary ions in TIC and ICR spectra are related by [114]

$$\frac{I_1}{I_2} = \frac{R_1}{R_2} = \frac{A_1}{A_2} \frac{m_2}{m_1}.$$

C. Apparatus

Crossed electric and magnetic fields have been used for mass analysis in several mass spectrometers but do not have wide application despite their perfect double focusing. The first cycloidal mass spectrometer was constructed in 1938 by Bleakney and Hipple [1]. A commercial instrument was described by Robinson and Hall [11], and some years ago the M 66 mass spectrometer was available for a short time from Varian Associates. Goudsmit [3] proposed a time-of-flight mass spectrometer using crossed electric and magnetic fields. The correct pitch of the helical ion path to reach the collector could be obtained with a controllable vertical deflecting field and a vertical collimator near the ion source. An improved version of this apparatus using a time-dependent electric field was constructed by Hipple and Thomas [4].

The ICR principle was first applied by Hipple et al. [5] who developed the omegatron for ionic mass determination. Ions are collected at an electrode which they strike only at resonance because of their larger orbits.

The first ICR spectrometer for study of ion–molecule collisions was developed by Wobschall et al. [18,21,23]. A solenoid is used to produce the magnetic field and ions drift along the solenoid axis to collide with a neutral gas. An rf field is produced by a pair of electrodes which form one arm of an rf bridge, and ions are detected at resonance by the imbalance they create in the rf bridge circuit. The disadvantage of this design lies in the difficulty of constructing a high-field solenoid with homogeneity over a large volume.

An ICR spectrometer designed by Llewellyn [132,132a,133], Baldeschwieler [87] and Beauchamp [88] has become commercially available from Varian Associates [20] and is now in widespread use. It is called "Syrotron" from the Greek for "to sweep". Since the majority of results discussed in the present article are derived from investigations with this instrument, it will be described in some detail.

In the standard model a three-section cell, shown schematically in Fig. 1, is mounted in a vacuum chamber and placed between the pole caps of an electromagnet. A Hall effect sensor probe regulates the intensity of the magnetic field and modulation can be produced by additional Helmholtz coils. A sweep control unit serves to sweep either magnetic field strength, electron energy, or the frequency of an rf oscillator. Several electronic units supply the voltages (drift, trapping, electron-accelerating) needed to make the cell function. Electron energy and emission current are regulated for constant ion production. Energy absorption by resonant ions is recorded by a marginal oscillator whose output is amplified and fed to a phase-sensitive detector. This detector is linked to a reference modulation oscillator, detects and processes signals possessing the modulation frequency, and has a dc output signal which is plotted on an x-y recorder.

The sensitivity of marginal oscillator detectors, on which the sensitivity of signal detection in an ion cyclotron resonance spectrometer depends, was originally discussed by Anders [26]. Corresponding discussions of the application of marginal oscillators in NMR spectrometry can be found in the literature [11,147]. The effective noise voltage V_n for the tank circuit of the oscillating detector for a noise level of unity is given by:

$$V_n = (4kTBQL^{1/2}C^{-1/2})^{1/2}$$

where T is the temperature (°K), B the bandwidth, Q the quality factor of the circuit, L the inductivity and C the capacitance. With typical values for an ICR circuit ($L = 5.56$ mH, $C = 50$ pF, $B = 10^{-1}$ sec, $T = 300$ °K, $Q = 300$) Beauchamp [147] calculated a noise level of 7×10^{-7} V, so that the marginal detector can be used to detect as few as 14 ions in the cell.

Beauchamp [147] and Drewery, Goode and Jennings [202] have estimated the resolving power of an ICR spectrometer in the limiting case of low pressure. At high magnetic field strengths (10 kG) a resolution of approximately 1000 can be obtained for $m/e = 100$.

With magnetic fields up to 14 kG and marginal oscillator frequencies down to 100 kHz, only a limited mass range to about 200 mu can be determined. In the observation of ion–molecule reactions the following characteristics are used:

1. Long ion path, hence high collision number, and secondary ion yield in the analyzer cell at low pressures (10^{-6} Torr).

2. Ion production, reactions with molecules, and detection of ions in separated regions.

3. The electric fields used being very small (*e.g.* 10 mV cm^{-1}), the ions drift with nearly thermal velocity (20 m sec^{-1}) on a cycloidal path.

4. Ions of specific charge: mass ratio can be detected and accelerated by application of resonant rf fields.

D. Measuring Techniques

In conjunction with the sensitive lock-in amplification and signal detection various modulation schemes are used which involve either a disturbance of the resonance state or a change in the density of the ion swarm drifting through the cell.

One scheme involves superimposing a *modulated magnetic field* on the static magnetic field, thus periodically changing the cyclotron frequency of the ions. As the frequency of the oscillator circuit is fixed, periodic resonance in power absorption occurs; this can be subjected to phase-sensitive detection. The same general type of signal (a derivative of the absorption signal) is obtained when the magnetic field is held constant and the frequency of the marginal oscillator is modulated.

Another scheme involves *modulation of the drift voltages* which, by varying the ion transit time, produces modulations in the density of ions in the cell; the marginal oscillator records a periodic change in energy absorption which is detected as an absorption signal at the modulation frequency.

The ion density can also be changed by periodically interrupting ion production with constant drift. Henis and Frasure [48] used *electron energy modulation* in which ion current production is interrupted by switching the electron energy above and below the ionization threshold. This technique can be useful for regulating specific ion species in the swarm. Electron energy modulation is satisfactory for cations, but not for anions. To obtain a mass spectrum at 70 eV, this scheme requires a high pulse amplitude which can cause excessively noisy signals due to pickup on the marginal oscillator.

McIver [139] suggested a *pulsed-grid modulation scheme* for modulating ion density. By suitably pulsing the bias on a grid between filament and trapping plate, it is possible to admit or block the entry of electrons to the cell. For emission currents below $3 \mu A$, a potential barrier of 1 V above the dc level of the filament satisfactorily blocks the flow of electrons into the ionization region. Modulation of the electron beam current generally works equally well for both positive and negative ions.

As McMahon and Beauchamp [182] have shown, it is possible to record ICR single-resonance spectra by *modulating the trapping voltage*, provided one trapping plate is pulsed between a positive and a negative potential with a period considerably longer than the ion transit time. A square wave voltage is applied to one trapping plate, switched between the levels $\pm v_T$. The potential applied to the opposite trapping plate is $+ v_T$ for positive ions and $- v_T$ for negative ions.

If the trapping potential is changed periodically with a definite frequency, the *ion ejection technique* developed by Beauchamp and Armstrong [58] can be used. In addition to cycloidal motion perpendicular to the magnetic field, ions undergo simple harmonic motion parallel to it in the presence of the trapping field caused by the trapping potential v_T. The frequency of this oscillation is

$$\omega_T = \sqrt{\frac{4 v_T q}{m d^2}}$$

where d is the distance between the trapping plates. The amplitude of these oscillations can be altered by applying an rf field of frequency ω_T to the trapping plates. Ions of given m/q are ejected from the ion swarm. Because ω_T is independent of magnetic field strength, spectra without the ejected ions can be obtained normally. However, resolution is low, which means that usually more than one ion species is ejected. The requirements of efficient ion ejection and high mass resolution are conflicting. For this reason Dunbar [153] used a technique in which the ions are ejected by means of an rf field in the plane of their cyclotron motion. As with the TIC technique, the ion cyclotron orbits are increased in radius by the ejection rf field at resonance frequency until the ions strike the drift plates and are removed from the cell. The amplitude of the ejecting field must be large enough to ensure that ions are ejected soon after their arrival in the cell. ICR ejection appears to be the method of choice in cases where relatively long ejection times can be tolerated but high mass resolution is essential [153].

Another technique which has no parallel in conventional mass spectroscopy and thus underlines the usefulness of ICR spectroscopy is the *double-resonance technique* [24]. In most cases the yield of an ion–molecule reaction

$$P^+ + N \longrightarrow S^+ + R$$

varies with the translational energy of the reactant ion P^+. This fact is used to obtain ions which are coupled by IMR. The magnetic field is adjusted until S^+ ions absorb energy from the observing oscillator of frequency ω_S; the signal intensity of S^+ is monitored. If a second rf field is now applied to the source region of the cell and the frequency of the oscillator, ω_P, equals the cyclotron frequency of P^+, these ions absorb energy so that their translational energy increases and a change in the S^+ intensity is observed. At a fixed magnetic field strength the following relation holds:

$$\frac{m_P}{m_S} = \frac{\omega_S}{\omega_P}.$$

Thus, by scanning ω_P and monitoring the intensity of S^+, different precursors of S^+ can readily be identified from the changes in the intensity of S^+ at the double resonances. The direction of the change (increasing or decreasing intensity) shows the dependence of the ion–molecule reaction on kinetic energy.

Several special ICDR techniques have been applied.

1. A substantial improvement in sensitivity was obtained by Llewellyn [132a] who connected the irradiating transmitter and the marginal oscillator in series to the opposite drift plates of the analyzer region.

2. If the irradiating transmitter is connected to the source region and the oscillating detector to the analyzer region, unimolecular dissociation mechanisms may be identified.

3. In the pulsed double-resonance method [24], the rf oscillator is modulated and from the phase-sensitive detector a signal is obtained which is proportional to the difference between the spectra with and without double resonance.

An ICR spectrometer with an optical glass window fitted to the end of the vacuum can [133] is suitable for observing the *influence of photons* upon ions or ion–molecule reactions. Ions produced by electron impact can be trapped in the analyzer cell for periods in the range from msec up to 1 sec [153]. A light beam is passed through the window along the longitudinal axis of the cell, collinear with the drift path of the ions. The ions of interest are usually observed by monitoring power absorption by means of a marginal oscillator tuned to the cyclotron frequency. The change in ion intensity can be plotted against the wavelenght of the light. An analysis of the curve representing experimental observations can provide data of interest (like electron affinities, or dissociation energies) or give an insight into the reaction mechanism. This technique has been used to study photodissociation, photodetachment, and photon-induced IMR.

Types of Cells

Standard three-section cells are available in square (2.54 × 2.54 × 12.7 cm) or flat form (1.27 × 2.54 × 12.7 cm). For the study of special collision problems some advantage has been achieved by using ICR cells of special design in conjunction with special pulsing techniques.

Clow and Futrell [107] use a 12-in. magnet which permits the use of a longer ICR cell with an additional reaction zone between the source and analyzer regions. The source region has separate trapping plates from those of the reaction-analyzer zones. Better ion transmission is achieved by using higher trapping fields in the reaction-analyzer regions, whereby the ion motion is confined to the center of the cell. All drift and trapping potentials can be varied independently; a particular feature is that ion ejection in the source region is possible. Typical with this technique is a pulse sequence comprising ion creation, excitation, reaction and detection. An ion bunch is created by pulsing the electron energy from just below the ionization potential to a few volts above it at a frequency of 1 kHz for a duration of 100 μsec. In the following 250 μsec the irradiating rf oscillator is activated at the cyclotron resonance frequency of the ions to bring them to a definite energy. Subsequent reaction with neutrals occurs as the ions drift through the reaction zone. A static magnetic field is used and the marginal oscillator is swept through the desired resonance frequencies. Currents are then measured using the ion collector. The authors also use a modification of this technique: a pulsed double-resonance scheme in which the phase-sensitive detector amplifies the difference in signal level for selected product ions produced alternately by thermal energy and translationally excited ions. Furthermore [200] they tried to increase ion transmission by introducing specially shaped reaction-analyzer drift plates: to provide a more uniform electric field in the center of the cell, the outer one-fourth of the drift plates was inclined 50° from the horizontal towards the center.

Huntress [214] used a modified version of the four-section cell without ion-collecting region and having a short distance (1 cm) between electron beam and end of the source region.

Another modified four-section cell constructed by Marx and Mauclaire [257] has reaction-analyzer drift plates extending to the ion collector region.

McIver [138] replaced the normal three-region cell by a single-region cell (2.54 × 2.54 × 8.9 cm). His "trapped-ion analyzer cell" differs in three respects from the standard cell: firstly gaseous ions are produced and detected in the same region; secondly, his cell has plates at both ends to trap the ions inside it; and thirdly, the trapped-ion analyzer cell relies entirely on pulsed modes of operation. The side plates provide potentials slightly more negative than the trapping potential. Observing and heating rf fields are coupled in the normal way to the drift plates. Positive ions are trapped in the cell by biasing the drift plates and the two end plates equally at a voltage from 0.5 to 3 V more negative than the trapping voltage. Negative ions are trapped by simply reversing the polarity of the dc voltages on the cell plates. An experiment is initiated by a 0.10 msec electron-beam pulse controlled by pulsed grid modulation [139]. In the following time τ_1 the product ions are allowed to interact with neutrals, either with their thermal energy or with increased kinetic energy caused by an rf pulse of suitable resonance frequency. After the desired reaction time the ions are detected during the time τ_2 by pulsing the intensity of the magnetic field so as to satisfy the resonance condition between marginal oscillator frequency and cyclotron frequency. The detection period is followed by a quenching period: here the polarity of the dc voltage of the upper drift plate is temporarily inverted which destroys the trapping action of the cell, and all ions, whatever their mass, drift to the walls. This allows a high repetition rate of sequences and prevents overlapping of ions from earlier sequences into later ones. The trapped-ion technique has been improved [180] for pulsed double-resonance experiments: instead of pulsing the magnetic field for detection to maintain the resonance condition at a fixed marginal-oscillator frequency, the shift in cyclotron frequency due to the applied trapping voltage is compensated for by an increase in the magnetic field. The required increase can be calculated. A special pulse-compensator circuit was developed to reduce coupling effects of the irradiating and marginal oscillators. The trapped-ion technique should be useful for studying the dependence of ion molecule reactions on kinetic energy up to 40 eV.

Studies undertaken by means of analysis of the fields in ICR cells have given some insight into the real motion of the ions and how this depends on the different parameters [182,231]. In particular, the effect of quadrupolar electric fields on ion motion under ideal conditions has been studied by Sharp et al.[231]. Potentials measured inside the cell enable ion motion to be predicted.

II. Line Shape and Rate Constants

A. Power Absorption and Line Shape

TIC spectra can be interpreted in a straightforward manner whereas ICR spectra involve some critical factors as regards both obtaining and interpreting them. Measuring the power absorption from the rf field provides information in terms of peak height and peak width about the ion system represented. It is known from experiment that linewidth and line shape for a given reactant or product ion depend on lifetime and collision frequency. Hence for quantitative investigations a detailed analysis of the ICR signal line shape is desirable.

Wobschall et al.[18] were the first to give an expression for ICR power absorption in their treatment of collision-broadened lines. A series of papers which appeared in the Journal of Chemical Physics between 1967 and 1971 dealt with the analysis of expressions of energy absorption for different ion–molecule reactions. More recently Comisarow [152] put forward a comprehensive theory of ICR power absorption.

H. Hartmann, K.-H. Lebert, and K.-P. Wanczek

To understand the information conveyed by an ICR signal, one has to solve the equation of motion of an ion under ICR conditions. The effect of collisions on the energy absorption of ions in the ICR cell is described by the Lorentz equation with the inclusion of a term for collisional damping:

$$\bar{v} = \frac{q\bar{E}}{m} + \left[\frac{q}{m} \, \bar{v} \times \bar{B} \right] - c \, \bar{v} \tag{5}$$

where $\bar{v}$ is the average velocity of an ion of mass m and charge q, $\bar{E}$ is the electric field, $\bar{B}$ the magnetic field, and c is the reduced collision frequency, which is independent of the average ion velocity. This independence applies only when the ion–neutral interaction potential depends upon the inverse fourth power of the ion–neutral distance. This holds for the ion-induced dipole potential.

If the linearly polarized electric field E is decomposed into a sum of two counter-rotating circularly polarized fields

$$E(t) = E_1 \cos \omega t \bm{j} = E^+ + E^-$$
$$E^+ = \tfrac{1}{2} E_1 \sin \omega t \bm{i} + \tfrac{1}{2} E_1 \cos \omega t \bm{j}$$
$$E^- = -\tfrac{1}{2} E_1 \sin \omega t \bm{i} + \tfrac{1}{2} E_1 \cos \omega t \bm{j}$$

where E_1 is the peak value of the rf field. The components of Eq. (5) for a cation are given by

$$v_x = (qE_1/2m) \sin \omega t + \omega_c v_y - c v_x$$
$$v_y = (qE_1/2m) \cos \omega t - \omega_c v_x - c v_y$$

where ω_c is the cyclotron frequency (1). The components of velocity are

$$v_x = B \cos \omega t + C \sin \omega t + (D \cos \omega_c t + F \sin \omega_c t) \exp(-ct)$$
$$v_y = C \cos \omega t - B \sin \omega t + (F \cos \omega_c t - D \sin \omega_c t) \exp(-ct)$$

where

$$B = \frac{qE_1[(c^2 + \omega_c^2 - \omega^2)(\omega_c + \omega) - 2c^2\omega]}{2m[(c^2 + \omega_c^2 - \omega^2)^2 + 4c^2\omega^2]}$$

$$C = \frac{qE_1 c[2\omega(\omega_c + \omega) + c^2 + \omega_c^2 - \omega^2]}{2m[c^2 + \omega_c^2 - \omega^2)^2 + 4c^2\omega^2]}$$

$$D = v_0 \sin \gamma - B; \quad F = v_0 \cos \gamma - C$$

68

t is the time during which the rf field was applied, γ is the phase angle of initial velocity. From the electric field E the power

$$A = qE(t)(v_x + v_y)$$

is absorbed. Thus, the instantaneous power absorption is

$$A(t,\omega_c,\omega,c,\gamma) = \tfrac{1}{2}E_1 q\{[-D\sin(\omega_c - \omega)t + F\cos(\omega_c - \omega)t] \times$$
$$\exp(-ct) + C\}.$$

Since all values of the phase angle are equally probable, averaging over γ leads to:

$$A(t,\omega_c,\omega.c) = \frac{q^2 E_1^2}{4m(c^2 + w^2)}[(w\sin wt - c\cos wt)e^{-ct} + c] \qquad (6)$$

where t is the time the ion has been absorbing energy from the alternating electric field of peak value E_1 and frequency ω, and $w = (\omega_c - \omega)$. In the case of resonance, which is what mainly interests us, Eq. (6) becomes simpler

$$A(t,\omega_c = \omega,c) = \frac{q^2 E_1^2}{4mc}[1 - e^{-ct}]. \qquad (7)$$

This means that for short times the power absorption increases linearly with time and is nearly independent of c, while for long times it becomes constant over time. In the limit $c \to 0$ the "zero pressure" power absorption, also found by Buttrill [67] (excepted for an error by a factor of two in Buttrill's analysis), gives

$$A(t,\omega_c,\omega) = \frac{q^2 E_1^2}{4m(\omega_c - \omega)}\sin(\omega_c - \omega)t. \qquad (8)$$

With the approximation $c/t \gg 1$ an expression for high pressures is obtained from (6)

$$A(\omega_c,\omega,c) = \frac{q^2 E_1^2 c}{4m[c^2 + (\omega_c - \omega)^2]}. \qquad (9)$$

This had been found earlier by Wobschall et al. [18] and by Beauchamp [28]. Line shape expressions are obtained by summing the power absorption of all ions over the analyzer region

$$A_{total} = \int_0^\tau I(t) \cdot A(t)\,dt. \qquad (9a)$$

69

Expressions of power absorption are also obtained for systems in which ion–molecule reactions occur. Comisarow's theory involves two types of collision frequencies: the reduced nonreactive collision frequency c, and the chemically reactive collision frequency k, the first-order reaction rate of an IMR. c is introduced into the equation of motion (5) from which A is derived. k is introduced into a kinetic equation which gives the ion currents for primary, secondary and tertiary ions as a function of time. The calculation of the total power absorption in the case of IMR is a counting procedure according to Eq. (9a) which sums all the ions produced in all cell regions and all the power absorptions.

B. Rate Constants

Using Eq. (6) one obtains rather complex expressions for the total power absorption of primary, secondary, and tertiary ions. Unfortunately, these expressions cannot be readily rearranged for k, so a computer must be used to solve them iteratively for the reaction rate constant.

A simpler expression for the reaction rate constant in the important case of an IMR which leads to only one secondary ion species has been calculated by Buttrill [67] using power absorption (8) in the resonance case $A = q^2 E_1^2 t / 4m$

$$k = 3 m_p^2 I_s [nm_s^2 I_p (2\tau_p + \tau_{p'}) + nm_p^2 (\tau_p + 2\tau_{p'})]^{-1} \qquad (10)$$

where m_p, m_s, I_p and I_s are the mass and single-resonance intensity of the primary and secondary ions, and τ_p and $\tau_{p'}$ are the times at which the primary and secondary ions enter and leave the resonance region. There is an extension to this formula, given by Marshall and Buttrill [137], for IMR in which tertiary ions are produced.

A similar analysis to derive an expression for k is used by Goode et al.[114] which leads to

$$k = m_p^2 I_s [n (m_s^2 I_p + m_p^2 I_s) (\tau_{p'} + 2\tau_p)/3]^{-1}. \qquad (11)$$

The difference between expressions (10) and (11) is due to the different approximation procedures used. In both cases the signal intensities for primary (I_p) and secondary (I_s) ions are calculated to determine k, but an analytical expression for k can only be obtained by an approximation of the exponential functions. Buttrill calculates the exact expressions for I_p and I_s by integration and approximates the remaining exponentials. Goode et al. integrate over approximated ion currents.

Thus, slightly lower k values are obtained if Eq. (11) is used instead of (10).

Another type of analysis is used by Bowers *et al.*[40] for the case of collision-limited power absorption by the reactand and product ions. The rate constant for IMR producing only one secondary ion species in the limit of low conversion is given by

$$k = 2m_p \xi_s I_s [n(m_p \xi_s I_s + m_s \xi_p I_p)(\tau_p + \tau_{\acute{p}})]^{-1} \qquad (12)$$

where ξ_p and ξ_s are the collision frequencies for momentum transfer of the primary and secondary ions, respectively. Again, the rate constants obtained by Eq. (11) are lower than those by Eq. (12), as McAllister[226] has shown for a special reaction.

All ion–molecule collisions reduce ICR signal intensity (peak height) since they interfere with energy absorption. Reactive collisions are more effective in this respect than nonreactive ones, because they change the mass of the absorbing ion and completely prevent subsequent energy absorption[137]. Consequently, if a sequence of ion–molecule reactions $(P^+ \rightarrow S^+ \rightarrow T^+)$ is observed, the changes in signal intensity differ according to whether the products are stable or not.

At present, easily handled expressions which allow rapid determination of rate constants are available only for the limiting case of IMR systems with low conversion. Expressions for k derived from "zero-pressure" power absorptions are systematically too low. They decrease with increasing pressure and increasing residence time[152]. On the other hand, it does not seem entirely clear how to establish experimental ICR parameters which fulfill the conditions required by the theory[226]. Most of the experimental uncertainties stem from pressure measurements and from determination of ion transit times in the cell. Capacitance manometers and four-sectioned cells in connection with a drift pulse technique[114] seem better experimental equipment for the purpose of measuring rate constants of IMR. For future development the trapped-ion cell developed by McIver[138,180] promises some advances. The present "absolute" rate constants of IMR must be treated as estimates of poor accuracy.

To study the dependence of IMR upon kinetic energy, the reaction yield is observed while the primary ion is being heated to higher kinetic energies by an rf field at resonance. The excess over thermal energy of an ion spending time t in an rf field of strength E_1 was calculated by Buttrill[67]:

$$I = [q^2 E_1^2 / 2m(\omega_1 - \omega_c)^2] \sin^2[\tfrac{1}{2}(\omega_1 - \omega_c)t]. \qquad (13)$$

71

Somewhat different expressions were obtained by Bowers *et al.*[40] and by Huntress [171] but in the case of resonance $(\omega_1 = \omega_c)$ from all expressions the average energy absorbed by an ion is given by

$$T = \frac{qE_1^2\tau^2}{8m} \tag{14}$$

where τ is the time the ion spends in the rf field. At resonance the kinetic energy of an ion depends on the square of the time during which it was absorbing energy. Thus τ must be known exactly.

The resonance absorption of an ion in an ICR cell can be treated as limited in either time or space. Therefore the duration of a pulsed rf field or the residence time in a continuous rf field must be known in order to calculate T. In pulsed ICDR experiments [55] the length of time the rf field is applied and the peak strength E_1 together determine the kinetic energy [29].

τ can be calculated from the geometry of the cell and the drift velocity, but ion residence time in an ICR cell is a complex function of many variables. As Smith and Futrell [262] have shown experimentally, calculations of residence times are correct only for very low electron currents and trapping voltages. When dc electron currents approach 0.5 μA, or when the trapping voltage is equal to or greater than one-half of the drift voltage, the theoretically calculated residence time may be in error by more than 25%. These authors therefore determine the residence time experimentally from the lag time of a pulse sequence and measurements of TIC [200,262].

III. Applications

A. Cross Sections and Collision Frequencies

The early ICR work of Wobschall and coworkers [16–18,22,34,35] included measurements of ion–molecule collision cross sections. In the high-pressure limit the half-width at half-maximum of a Lorentzian ICR lineshape is simply related to the reduced collision frequency ν_0:

$$\nu_0 = \Delta\omega_{1/2}$$

and the reduced collision frequency is related to the collision cross-section:

$$\sigma = \nu_0 \left(\frac{m+M}{m}\right)^{\frac{1}{2}} \frac{1}{n\bar{v}}.$$

There are two cases: low E/p and high E/p (E is the electric field strength and p the pressure), in which the energy gained between collisions is smaller or greater, respectively.

The collision cross section as a function of E/p is determined from the increase of linewidth with pressure. At higher E/p the hard-sphere collision model was found to be a good approximation. The main contribution to the cross section of ions in their parent gases is from charge transfer [35].

Dunbar [155], using his transient ICR technique, has determined the momentum relaxation rate constants for N_2^+/N_2, $k_r = 0{,}9 \pm 0{,}1$ cm^3 molecule^{-1} sec^{-1} and CH_4^+/CH_4, $k_r = 1{,}25 \pm 0{,}15$ cm^3 molecule^{-1} sec^{-1} from a plot of relaxation rate against pressure. A comparison of observed and predicted rates shows that N_2^+ has a substantial nonorbiting charge-transfer rate in contrast to CH_4^+. This leads to the suggestion that the geometry of CH_4^+ is very different from that of the neutral methane molecule. The nonorbiting charge-transfer cross section of methane is currently under study [233].

Huntress [171] obtained the most accurate values for momentum-transfer rate constants now available, by measuring the instantaneous power absorption in the low pressure limit ($E/p = 0$). Values obtained more recently [218] by the phase-coherent pulsed ICR method for momentum-transfer rate constants for the systems CO_2^+/CO_2 and N_2^+/N_2 are in good agreement with Huntress' measurements.

Beauchamp [147] and Kevan and Futrell [216] have measured mobility and collision rate constants

$$k_c = \frac{c}{n}$$

for several hydrocarbon and perfluorocarbon ions in their parent gases and for H_3O^+ in H_2 and N_2H^+ in H_2. It has been shown [216] that angle-averaged molecular polarizabilities can be calculated from ICR line broadening. However, in many cases the polarization potential is an inaccurate description for ion-neutral distances appearing during thermal-energy collisions (e.g. for H^+, H_3^+ in H_2).

B. Nucleophilic Displacement Reactions and Affinities

Beauchamp and coworkers [124] have reported the general occurrence of nucleophilic displacement reactions in the gas phase:

$$CH_3M^+ + N \longrightarrow CH_3N^+ + M. \qquad (R\,1)$$

Table 1. Collision cross sections, collision rate constants and mobilities

System	Cross section (Å)		Collision rate constant	Mobility (cm²V^{-1}s^{-1})	Ref.
	low E/p	high E/p	$c/n = k$, 10^9 (cm³molec.$^{-1}$s^{-1})		
N$_2^+$ in N$_2$	185 ± 30	125 ± 30	0.9 ± 0.1; 0.67 ± 0.01	1.5; 1.21 ± 0.01	18,155,171)
	250 ± 40				34)
Ar$^+$ in Ar	215 ± 40	140 ± 40		1.4	18)
	260 ± 40				34)
O$_2^+$ in O$_2$	150	70 ± 30		2.0	18)
	195 ± 30				34)
H$^+$ in H$_2$	135 ± 20		23.5 ± 0.6	14.7; 15.1 ± 0.4	34,147)
H$_2^+$ in H$_2$	275 ± 30			5.7	34)
H$_3^+$ in H$_2$	190 ± 20		11.59 ± 0.11; 10.9 ± 0.2	7.2; 11.22 ± 0.33	34,147,171)
H$_3^+$ in He	21 ± 10			55	34)
He$^+$ in He	125 ± 15			9.1	34)
O$_2^-$ in O$_2$	185 ± 40			2.1	34)
O$^-$ in O$_2$	110 ± 30			4.1	34)

CO_2^+ in CO_2	0.67 ± 0.01		1.21 ± 0.02		171)
CH_5^+ in H_2	1.82 ± 0.01		11.55 ± 0.04		147)
H_3O^+ in H_2	1.67 ± 0.03		11.25 ± 0.19		147)
N_2H^+ in H_2	1.10 ± 0.01		11.26 ± 0.10		147)
$C_2H_5^+$ in H_2	1.04 ± 0.02		11.88 ± 0.20		147)
$C_3H_7^+$ in H_2	0.73 ± 0.01		11.37 ± 0.07		147)
CH_5^+ in CH_4	8.90 ± 0.06	7.20 ± 0.7	2.36 ± 0.02	2.92	147,216)
$C_2H_5^+$ in CH_4	5.30 ± 0.12	4.43 ± 0.4	2.33 ± 0.06	2.79	147,216)
$C_3H_7^+$ in CH_4	4.11 ± 0.05		2.02 ± 0.03		147)
$C_4H_9^+$ in CH_4	3.12 ± 0.09		2.01 ± 0.07		147)
$C_3H_7^+$ in C_3H_8		6.90 ± 0.7		1.21	216)
$C_4H_9^+$ in $n-C_4H_{10}$		6.90 ± 0.7		0.91	216)
CF_3^+ in CF_4		3.77 ± 0.4		1.38	216)
$C_2F_5^+$ in C_2F_6		4.08 ± 0.4		0.74	216)
$C_3F_7^+$ in C_3F_8		3.90 ± 0.4		0.54	216)
$C_4F_7^+$ in $c-C_4F_8$		4.23 ± 0.4		0.47	216)

S_n reactions occur if

1. the reaction is exothermic (as is generally necessary for ion–molecule reactions) and

2. proton transfer from the protonated substrate to the nucleophile is endothermic.

Such reactions take place in order of decreasing methyl-cation affinity:

$$NH_3 > CO > H_2S > CH_2O > HI > H_2O > HBr > HCl > N_2 > HF.$$

Experimental determination will allow a quantitative assignment of "hardness" and "softness" of bases and acids. The difference in behavior towards a given neutral M of the "soft" acid CH_3^+ and the "hard" acid H^+ is determined by the difference in their methyl-cation and proton affinities.

Like the later discussed [246] formation of alkylfluoronium from protonated alkylfluorides, this reaction offers a novel means of nitrogen fixation [168]. The methyldiazonium ion in a mixture of $CH_3F:N_2:H_2 = 1:110:28$ makes up 43% of the total ion current at higher pressures

$$CH_3FH^+ + N_2 \longrightarrow CH_3N_2^+ + HF \tag{R2}$$

CH_3N^+ undergoes the nucleophilic displacement reaction with ammonia:

$$CH_3N_2^+ + NH_3 \xrightarrow{\hspace{2cm}} \begin{array}{c} CH_3NH_3^+ + N_2 \\ \cancel{\hspace{1cm}} \\ NH_4^+ + CH_2N_2 \end{array} \tag{R3}$$

but, as expected, no proton transfer is observed.

Using CO as a nucleophile, the acetyl cation is formed

$$CH_3FH^+ + CO \longrightarrow CH_3CO^+ + HF.$$

In the same way methylxenonium [167] is generated

$$CH_3FH^+ + Xe \longrightarrow CH_3Xe^+ + HF$$

but no reaction is observed with Kr. This indicates that the proton affinity of CH_3F is greater than the proton affinity of Kr.

Determination of proton affinities (basicities) and acidities in the gas phase provides a means of systematically representing a large number of ion–molecule reactions and of the relationships between these quantities and bond strengths, ionization potentials and electron affinities without disturbance by solvation phenomena [147].

The enthalpy change during the reaction

$$MH \longrightarrow M^- + H^+ \tag{R4}$$

defines the proton affinity $PA(M^-)$ of the anion M^- and is a quantitative measure of the acidity of the neutral MH. As can be shown by the corresponding thermodynamic cycle, $PA(M^-)$ is given by

$$PA(M^-) = D(MH) + IP(H) - EA(M)$$

where $D(MH)$ is the bond dissociation energy of MH to a radical and a H atom, $IP(H)$ is the ionization potential of H, and $EA(M)$ the electron affinity of the radical.

The hydrogen affinity $HA(M^+)$ of M^+, the enthalpy change for the reaction

$$MH^+ \longrightarrow M^+ + H, \tag{R5}$$

is given by

$$HA(M^+) = IP(M) - IP(H) - PA(M).$$

Relative proton affinities can be determined from proton transfer reactions

$$MH + N^- \longrightarrow M^- + NH \tag{R6}$$

and, in a corresponding manner, the acidities. Tables 2 and 3 list gas-phase acidities and basicities for a number of inorganic and organic compounds [147]. The relative proton affinities of several amines have also been determined [148,149,149b] but will not be discussed here.

In the case of the main-group hydrides, the basicity of the anions decreases in each group of the periodic system from lighter to heavier elements, whereas with the proton affinities of the neutrals a reversal is observed in the 7th group of the periodic system.

The anomalous order of amine basicities [192], observed in solution, is not found in the gas phase, as would be expected. The proton affinity of the methylamines increases with increasing methyl substitution:

$$(CH_3)_3N > (CH_3)_2NH > CH_3NH_2 > HN_3.$$

The hydrogen affinities of fluoromethanes increase with increasing fluorine substitution:

$$CH_4 < CH_3F = CH_2F_2 < CHF_3 < CF_4$$

but the proton affinity passes through a maximum:

$$CF_4 < CH_4 < CH_2F_2 = CHF_3 < CH_3F.$$

Nixon and Bursey [141] obtained the relative nitryl ion affinities:

$$HOH < \text{methanol} < \text{ethanol} < \text{i-propanol}.$$

The order of relative fluorine and chlorine affinities of halomethanes and haloethanes, which is a measure of the Lewis acidity, have also been investigated [247]. The order of increasing fluorine affinity is:

$$CH_3^+ > CF_3^+ > CHF_2^+ > CH_2F^+ > CHFCl^+ > CH_3CF_2^+ > CH_3CHF^+ > CH_3CH_2^+.$$

C. Photodetachment, Photodissociation, and Photon-Induced Reactions

Until recently very few photodetachment studies had been made of negative molecule ions. ICR methods [66,186,233,234,235] provide a direct measurement of the detachment process. If the force constants of ion and neutral are very similar, so as to involve no change in the vibrational quantum number during transition for the negative ion in the vibronic ground state to the neutral, the threshold at longest wavelength corresponds to the adiabatic electron affinity of the neutral.

With the aid of an extrapolation procedure, discussed in detail, Smyth and Brauman [233-235] have obtained the electron affinities of $NH_2\cdot$, $PH_2\cdot$, $AsH_2\cdot$, and $SeH\cdot$. These are shown in Table 4.

In the first three rows of the periodic system hydrides of group $-V$ and $-VI$ elements show the same behavior.

In the case of SeH^-, it was the first time molecular fine structure was measured by photodetachment [235]. Two thresholds assigned to transitions from SeH^- ($^1\Sigma^+$) to the two final states $^2\Pi_{3/2}$ and $^2\Pi_{1/2}$ lead to a determination of the molecular spin-orbit coupling constant: A ($SeH\cdot$) $= -1815 \pm 100 \text{ cm}^{-1}$.

However, the absolute photodetachment cross section is difficult to obtain without an exact knowledge of the time the ions spend in the light beam and the geometrical overlap factor.

Dunbar [153] has detected photodissociation of the CH_3Cl^+ and N_2O^+ cations. For the process:

$$CH_3Cl^+ \longrightarrow CH_3^+ + Cl \tag{R7}$$

Table 2. Gas-phase acidities and basicities of inorganic molecules [147]

	IP(M) (eV)	PA(M)	HA(M+)	Ref.	Bond strength	EA(radical)	PA(anion)	Ref.
SiH$_4$	11.80	146	105		80	24	369	201,233,264
NH$_3$	10.15	207	128	73,201,264)	109	17	405	201,233,264
PH$_3$	9.98	185	102	73,201,264)	84	29	369	201,233,264
AsH$_3$	10.03	175	92	201,264)	71	29	360	201,234,264
H$_2$O	12.60	164	143	39,201)	119	42	390	201)
H$_2$S	10.42	170	102	39,201)	90	53	350	201)
H$_2$Se	9.98	170	85	201)	76	50	339	201,235)
HF	15.77	131	182	201)	136	79	370	201)
HCl	12.74	140	121	201)	103	83	333	201)
HBr	11.62	141	96	201)	88	78	323	201)
HI	10.38	145	71	201)	71	71	313	201)
He	24.58	42	295					
Ne	21.56	48	235					
Ar	15.76	73.7; 82	132	95)				
Kr	14.00	103	112					
Xe	12.13	124	90					
CO		143		124)				
N$_2$		165		124)				
NF$_3$	12.95	151	138	169)				

All values are in kcal mole^{-1} and are taken from Beauchamp[147] unless otherwise stated.

Table 3. Gas-phase proton affinities of organic molecules [147]

	IP (M) (ev)	PA (M) kcal/mole	HA (M+) kcal/mole	Ref.
Methane	11.80	122	104	39,124)
Tetrafluoromethane	15.35	121	161	246)
Difluoromethane	12.72	147	127	246)
Trifluoromethane	13.8	147	152	246)
Fluoromethane	12.54	151	127	246)
Ethylene	10.45	160	87	39)
Acetylene	11.40	154	101	246)
		138	88	39)
Propylene	9.73	179	90	39)
trans-2-Butene	9.13	180	77	148)
cis-2-Butene	9.13	181	78	148)
1-Butene	9.58	183	90	
Isobutylene	9.23	195	92	39)
Methylamine	8.90	216.3	109	192,212)
Dimethylamine	8.21	222	99	192,212)
Trimethylamine	7.78	227	93	192,212)
Ethylamine	8.80	219	109	192,212)
iso-Propylamine	8.68	221	109	192,212)
tert-Butylamine	8.60	223	109	192,212)
Pyridine		225		238)
		225		169,192)
		227		169,192)
		228		169,192)
Acetonitrile	12.21	186	154	38)
Methylchloride	12.28	160	107	193)
Ethylchloride	10.97	167	107	193)
Methylbromide	10.53	163	93	193)
Ethylbromide	10.29	170	93	193)
Methyliodide	9.54	170	77	193)
Ethyliodide	9.33	175	77	193)
Dimethyl ether	10.00	186	103	
Diethyl ether	9.53	199	105	
Ethylene oxide	10.57	183	113	
Trimethylene oxide	9.85	190	104	
Methyl mercaptan	9.44	186	90	
Dimethylsulfide	8.69	197	84	

All values are taken from Beauchamp [147] unless otherwise stated.

he found a sharp maximum at 3150 Å with a photodissociation cross section of $7.8 \cdot 10^{-18}$ cm^2. The reaction

$$N_2O^+ \longrightarrow NO^+ + N \qquad (R8)$$

shows a cross section of $0.25 \cdot 10^{-18}$ cm^2 without a marked wavelength dependence in the range from 4000 to 6500 Å. However, the resolution was poor.

A gas-phase photon-induced ion–molecule reaction of the product ion $C_3H_5^+$ in the parent gas has been reported [217]:

$$C_3H_5^+ + C_2H_4 \rightarrow C_3H_7^+ + C_2H_2 \qquad (R9)$$

with an onset wavelength at 5200 Å and a yield increasing with wavelength.

Table 4. Electron affinities of group V and VI element hydrides

	Electron affinities (eV)	Ref.		Electron affinities (eV)	Ref.
OH ·	1.83 ± 0.04	[1]	NH$_2$ ·	0.744 + 0.022	[235]
SH ·	2.139 ± 0.01	[1]	PH$_2$ ·	1.25 ± 0.03	[234]
SeH ·	2.21 ± 0.03	[235]	AsH$_2$ ·	1.27 ± 0.03	[235]

[1] Literature cited in [235].

D. Ion–molecule Reactions of Hydrogen and Deuterium and Small Molecules of Atmospheric Interest

The ion–molecule reactions of H_2, D_2, and HD have been extensively studied:

$$H_2^+ + H_2 \longrightarrow H_3^+ + H \qquad (R10)$$
$$D_2^+ + D_2 \longrightarrow D_3^+ + D \qquad (R11)$$
$$HD^+ + HD \longrightarrow H_2D^+ + D \qquad (R12a)$$
$$\longrightarrow HD_2^+ + H \qquad (R12b)$$
$$H_2^+ + D_2 \longrightarrow H_2D^+ + D \qquad (R13a)$$
$$\longrightarrow HD_2^+ + H \qquad (R13b)$$
$$D_2^+ + H_2 \longrightarrow H_2D^+ + D \qquad (R14a)$$
$$\longrightarrow HD_2^+ + H \qquad (R14b)$$

Bowers, Elleman and King [62] have measured the rate constants and their dependence upon kinetic energy for the above reactions by the already described method. They showed that k_{10}, k_{11}, k_{12a}, k_{12b}, k_{13b} and k_{14b} increase initially with primary ion energy, reach a maximum and then decrease. Goode et al.[115] have called this energy dependence in question and attributed it to sweep-out and detuning effects. However, Clow and Futrell [200] confirmed the results of Bowers et al., except for k_{14b} which remains constant at low energies and decreases monotonically at higher energies. (The earlier statement that k was not dependent on energy was caused by greater experimental uncertainties). All reactions involving proton transfer show maxima in the kinetic energy dependence of rate constants. For reaction (R 10) Bowers *et al.* have proposed three mechanisms:

$$H_2^+ + H_2 \longrightarrow H_3^+ + H \quad \text{stripping}$$
$$H_2^+ + H_2 \longrightarrow (H_4^+) \longrightarrow H_3^+ + H \quad \text{complex}$$
$$H_3^+ \longrightarrow \text{products} \quad \text{destruction.}$$

The complex mechanism dominates at lower energy and the stripping mechanism at higher energy.

Under ICR conditions $(H,D)_3^+$ species are vibronically excited and there is a low threshold for H^+ and D^+ formation. A further value for k_{11} has been given by Inoue and Wexler [76]. The same authors showed the stripping mechanism to be appropriate to the detailed reaction mechanism of the nitrogen ion-hydrogen molecule reaction:

$$N_2^+ + HD \longrightarrow N_2H^+ + D$$
$$\longrightarrow N_2D^+ + H. \tag{R 15}$$

The reactions of H_2 and D_2 with rare gases [63,95,100] and with N_2O [259] have beeen studied. The detailed investigation of the ion–molecule reaction in the system $CO_2 - H_2$ has already been mentioned [262].

Reactions of hydrogen with small hydrocarbons like CH_4, CD_4 [76,94], C_2H_6 and C_2D_6 [189] show a great variety of product ions and a complicated isotopic exchange mechanism.

Vibrationally excited $(H_3^+)^*$ undergoes H^- abstraction reaction with ethylene oxide and acetaldehyde [147b] with a larger rate constant than collisionally deactivated H_3^+. From isotopic substitution studies a detailed reaction mechanism has been deduced:

(R 16)

E. Hydrocarbons

Although there are several studies of ion–molecule reactions of simple hydrocarbons, only the reactions of CH_3^+ and CH_4^+ ions of methane are known in any detail.

$$C^+ + CH_4 \longrightarrow C_2H_3^+ + H \qquad\qquad (R\,17)$$

$$CH^+ + CH_4 \longrightarrow C_2H_3^+ + H_2 \qquad\qquad (R\,18)$$

$$CH_2^+ + CH_4 \longrightarrow C_2H_2^+ + 2H_2 \qquad\qquad (R\,19a)$$
$$\longrightarrow C_2H_3^+ + H_2 + H \qquad\qquad (R\,19b)$$
$$\longrightarrow C_2H_4^+ + H_2 \qquad\qquad (R\,19c)$$

$$CH_3^+ + CH_4 \longrightarrow C_2H_3^+ + 2H_2 \qquad\qquad (R\,20a)$$
$$\longrightarrow C_2H_5^+ + H_2 \qquad\qquad (R\,20b)$$

$$CH_4^+ + CH_4 \longrightarrow CH_5^+ + CH_3 \qquad\qquad (R\,21a)$$
$$\longrightarrow C_2H_3^+ + 2H_2 + H \qquad\qquad (R\,21b)$$
$$\longrightarrow C_2H_4^+ + 2H_2 \qquad\qquad (R\,21c)$$

The ion–molecule reactions of methane [76,130] lead with the exception of the proton-transfer reaction (R21a) to a lengthening of the carbon chain, according to the general equation:

$$CH_n^+ + CH_4 \longrightarrow C_2H_{n+k}^+ + \frac{4-k}{2}\,H_2.$$

where $k \geqq 0$ except for (R21b).

The reaction channels (R21b) and (R21c) are possible only if the energy of CH_4^+ is greater than thermal [180]. CH_5^+ (R21a) mainly formed at thermal energies and only from CH_4^+ [243] begins to dissociate at higher energies [107,214] according to:

$$CH_5^+ \longrightarrow CH_3^+ + D_2. \qquad\qquad (R\,22)$$

The corresponding dissociation is observed also in the case of $C_2H_5^+$, raising the kinetic energy above 1 eV [55,107,214]:

$$C_2H_5^+ \longrightarrow C_2H_3^+ + H_2. \qquad\qquad (R\,23)$$

Reaction (R20b) has been studied by Huntress [214] using deuterated compounds. He showed, that complete equilibration of H and D atoms takes place in the reactions of CH_3^+ and CD_4 and CD_3^+ and CH_4, producing $C_2(H,D)_5^+$ in the energy range from thermal energy to 10 eV. No stripping mechanism was observed. The $C_2(H,D)_3^+$ ions formed at higher energy seem to be generated via an excited intermediate complex and not via a stripping mechanism.

The ion–molecule reactions of tetradeuteromethane investigated by Futrell [253] and of CH_4 show no substantial difference.

Clow and Futrell[68] have investigated charge exchange in the system $Xe - CH_4$. They showed mutual charge exchange between Xe and methane with the fortuitous result

$$Q_{Xe}k^{Xe^+,CH_4} = Q_{CH_4}k^{CH_4^+,Xe}$$

where Q is the ionization cross-section and k the bimolecular rate constant for charge transfer. No charge exchange is observed between Xe and CH_3.

Huntress and Elleman [125] have investigated the ion–molecule reactions of $N(H,D)_3$-$C(H,D)_4$ mixtures. They observed proton transfer charge transfer, hydrogen abstraction, and one condensation reaction

$$CH_3^+ + NH_3 \longrightarrow CH_4N^+ + H_2, \tag{R24}$$

leading to internally excited CH_4N^+. The latter will be discussed in some detail. The reaction proceeds by two mechanisms without the H-atom randomization observed for CH_4. The first mechanism results in H_2 loss across the new C–N bond producing the excited ion, the second in H_2 loss at the nitrogen, leading to CH_3NH^+:

$$CD_3^+ + NH_3 \longrightarrow (CD_3NH_3^+)* \nearrow \begin{array}{l} (CD_2 = NH_2^+) + HD\ 70\% \\ \searrow\ CD_3NH^+ \quad + H_2\ 30\% \end{array} \tag{R25}$$

The results may be compared with the ion–molecule reactions of $C_3H_6^+$ and NH_3 leading to $CH_2NH_3^+$ and $CH_2NH_2^+$ [207].

During the ion–molecule reactions of H_2S with C_2H_4 and C_2H_2 [106] complete H-atom scrambling is inhibited by the heteroatom, too. For both reactions a four-center mechanism can be written. The reaction of C_2D_4 with H_2S:

$$\tag{R26}$$

shows that all D atoms are retained at their carbon atoms.

The ion–molecule reactions of ethane [189,204,224] lead to a great variety of products. Depending on the internal excitation of $C_2H_6^+$, only one secondary ion with elongated carbon chain, $C_3H_9^+$, is observed at 70 eV electron energy [189], but at 13 eV [224] $C_3H_8^+$, $C_4H_9^+$, $C_3H_7^+$ and $C_4H_{10}^+$ could be detected, too. The protonated ethane, $C_2H_7^+$, formed with considerable excess energy according to

$$C_2H_6^+ + C_2H_6 \longrightarrow C_2H_7^+ + C_2H_5 + 25 \text{ kcal} \qquad (R27)$$

is no longer observed at 13 eV, but $C_3H_9^+$ can be detected:

$$C_2H_6^+ + C_2H_6 \longrightarrow C_3H_9^+ + CH_3 - 7 \text{ kcal.} \qquad (R28)$$

At pressures greater than 10^{-4} torr the probably collisionally stabilized ion $C_4H_{11}^+$ is found [224].

D labeling shows a different behavior of ethylene and its higher homologues. While complete randomization takes place in a mixture of C_2H_4 and C_2D_4 in the intermediate complex [40,214] (also in mixtures of C_2H_2 and C_2D_2 [40]), this is not the case for higher olefins [118] and mixtures of 2-C_4D_8 with various pentenes and hexenes [119]. This suggests quite specific structures for the intermediate complexes. However, products from the same intermediate can show different degrees of randomization.

The neutral 1-olefines are in general more reactive than 2-olefin or 3-olefin neutrals.

Almost all the ion–molecule reactions of acetylene [67,80] can be subsumed under the scheme:

$$X^+ + C_2H_2 \longrightarrow C_2HX^+ + H$$

where $X = C$, CH, C_2, C_2H and C_2H_2.

Isotopic mixing experiments with the isomeric compounds [100] methylacetylene (propyne) and allene showed that a four-center intermediate is important in many of their condensation reactions. The intermediate complex $(C_6H_4D_4^+)^*$ loses preferentially $C_2H_2D_2$

$$
\begin{array}{c}
[D_2C=C=CD_2]^+ \\
+ \\
H_2C=C=CH_2
\end{array}
\longrightarrow
\left[
\begin{array}{c}
D_2C\overset{\cdot}{=}C=CD_2 \\
\vdots \quad \vdots \\
H_2C\overset{\cdot}{=}C=CH_2
\end{array}
\right]^+
\longrightarrow
\begin{array}{c}
D_2C \\
\parallel \\
H_2C
\end{array}
+ C_2H_2D_2^+ \qquad (R29)
$$

In propyne $C_2H_2D_2$ is eliminated at approximately one-half the random prediction, and an intermediate may be:

$$\begin{bmatrix} H_3C-C\equiv CH \\ \vdots \quad \vdots \\ DC\equiv C-CD_3 \end{bmatrix}^+$$

Effects of nuclear transformation on ion–molecule reactions following β decay in tritiated cyclobutane and cyclopentane have been investigated by Pobo and Wexler [261]. β decay in tritium-labeled compounds RT gives a carbonium ion and a neutral helium atom:

$$RT \longrightarrow R^+ + He^3 + \beta^-. \tag{R30}$$

Double-resonance experiments with tritium-labeled cyclobutane show that the principal ion–molecule reactions are hydride and tritide transfer and a condensation reaction:

$$C_4H_nT_m^+ \ (n+m=7) \longrightarrow C_5H_nT_m \ (n+m=9). \tag{R31}$$

Only a condensation reaction

$$C_5H_nT_m \ (n+m=9) \longrightarrow C_6H_nT_m \ (n+m=9) \tag{R32}$$

has been evidenced for tritium-labeled cyclopentane, but no ICDR experiment could be performed.

Structurally different C_8H_8 radical cations are produced in the ICR spectra of styrene and cyclooctatetraene [190]. The styrene cations react with neutral styrene to produce a complex $C_{16}H_{16}^+$ (R33a) which loses benzene and forms $C_{10}H_{10}$ (R33b) via 1,4-elimination:

$$\tag{R33}$$

F. Structure Determination and Rearrangements

Jaffé and Billets [215)] have shown that the fragment ion $C_2D_3H_2^+$ formed from diethyl-N-nitrosamine-d_6 transfers only D^+ and not H^+ ions to the neutral nitrosamine. They concluded that $C_2D_3H_2^+$ exists without any H–D randomization as a localized $CD_3CH_2^+$ ion. The determination of the $C_3H_6^+$ ion structure by Gross [207)] and by Gross and McLafferty [181)] can be used as an example for one of the general methods. The $C_3H_6^+$ ions are generated from cyclopropane and from tetrahydrofurane. A labeled tetrahydrofurane-2,2,5,5-d_4 reacts to form $C_3D_4H_2^+$:

(R34)

The cations formed undergo reactions with added NH_3 or ND_3 and the distribution of product ions can be determined. If an acyclic $C_3H_4D_2^+$ cation is formed, either $CD_2NR_3^+$ and $CD_2NR_2^+$ ($R=H,D$) or $CH_2NR_3^+$ and $CH_2NR_2^+$ are possible as product ions, while in the case of a cyclic cation all four product ions should be detectable. The results indicate a cyclic structure and no randomization of the hydrogen atoms in the $C_3H_6^+$ cation or in the collision complex, and the following mechanism is deduced:

(R35)

In the case of the propyl ion, it could be shown that there is almost complete isomerization of n-$C_3H_7^+$ and cyclo-$C_3H_7^+$ to sec-$C_3H_7^+$ before ion–molecule reactions can take place [221)]. For protonated cyclo-propane, however, an alternative explanation is possible if $\Delta H_f($c-$C_3H_7^+)$ is higher than $\Delta H_f($sec-$C_3H_7^+)$.

Reactions of the 1,3-butadiene cation with various isomeric pentenes [164,207] show for each pentene a unique complex $(C_9H_{16}^+)^*$ which decomposes in a specific manner without the randomization observed in the case of small hydrocarbons. It is noteworthy that the isomeric cyclopentane does not react with $C_4H_6^+$.

Hoffman and Bursey [166)] proved the structure of the toluene molecular ion to be similar to that of the neutral. Ring expansion does not take place.

87

H. Hartmann, K.-H. Lebert, and K.-P. Wanczek

Three of the $C_2H_5O^+$ structural isomers [91]:

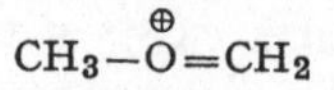

$$CH_3-\overset{\oplus}{O}=CH_2 \qquad CH_3-C\overset{\overset{\oplus}{O}H}{\underset{H}{\diagdown}} \qquad \overset{H}{\underset{CH_2-CH_2}{\overset{\oplus}{O}}}$$

1 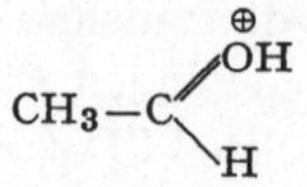   *2* 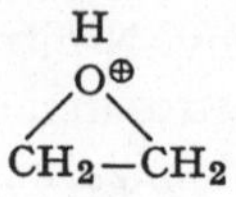   *3*

are commonly significant in ion–molecule reactions of this ion. In the ion–molecule reactions of *1* with methyl ethyl ether and of *2* and *3* with 2-propanol, only *1* can be distinguished from *2* and *3*. This may result from rearrangement of the protonated ethylene oxide *3* into protonated acetaldehyde *2*.

When investigating proton transfer in a series of dialkyl-N-nitros-amines [92], which requires an intramolecular rearrangement analogous to the first step in the McLafferty rearrangement:

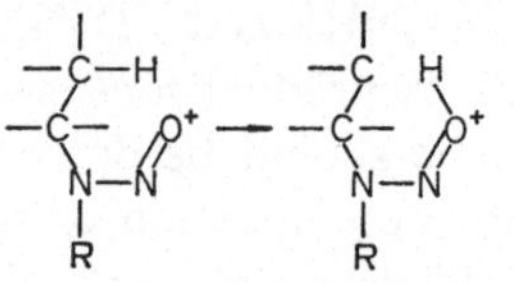

$$\begin{array}{ccc} -\overset{|}{C}-H & & -\overset{|}{C} \quad H \\ -\overset{|}{C}- \quad \diagup\!\!\overset{O^+}{} & \longrightarrow & -\overset{|}{C}- \quad \diagup\!\!\overset{O^+}{} \\ \overset{|}{N}-N & & \overset{|}{N}-N \\ R & & R \end{array} \qquad\qquad (R\,36)$$

for diethyl-N-nitrosamine the $C_2H_5^+$ fragment ion is observed.

The McLafferty rearrangement has been subject of several papers [69, 70, 110, 211].

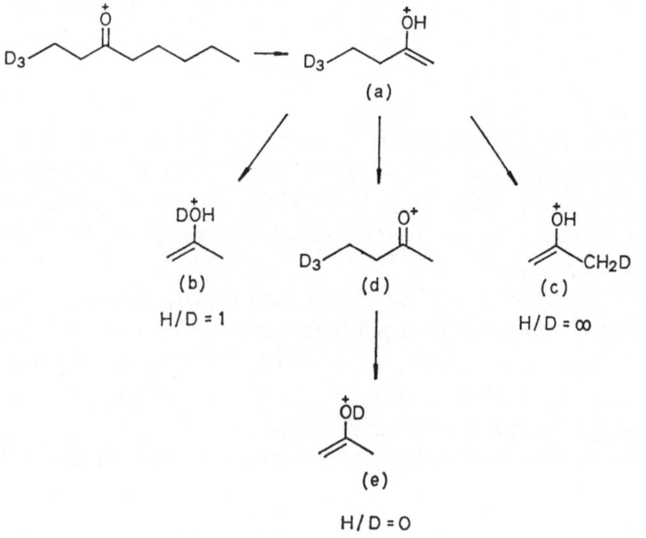

(R 37)

With several differently D-labeled 4-nonanones it could be shown [110] (as illustrated for 4-nonanone-1,1,1-d_3) that the double McLafferty rearrangement ion exists in an enolic form, which is created by hydrogen migration at the carbon–carbon double bond (c) and not by the isomerization of the oxonium ion (b) or by ketonization (e) of the intermediate single McLafferty enol. It is remarkable that no equivalent proton-transfer reaction could be observed for the dienol pent-1,3-dien-4-ol ion [219].

G. Haloalkanes and Haloalkenes

Ion–molecule reactions of fluoromethanes have been investigated by Marshall and Buttrill [137] (CH_3F) and by Beauchamp and coworkers [193, 246,260]. The ion–molecule reactions initiated by the parent ions of these compounds are systematic. The monofluorides and monochlorides [193] react to form the protonated parent ions:

$$RX^+ + RX \longrightarrow RXH^+ + R'X \qquad (R38)$$

from which subsequently the dialkylhalonium ion is formed:

$$RXH^+ + RX \longrightarrow RXR^+ + HX. \qquad (R39)$$

No hydrogen atom scrambling occurs, as can be shown by D-labeling:

$$CD_3^+ + CH_3F \longrightarrow CD_3\overset{+}{F}CH_3 \underset{47\%}{\overset{53\%}{\left\lceil \begin{array}{l} \longrightarrow CH_2F^+ + CD_3H \\ \longrightarrow CD_2F^+ + CH_3D \end{array} \right.}} \qquad (R40)$$

For the monobromides and monoiodides, however, no protonated parent ion occurs. At a relatively slow rate constant the dialkylhalonium ion is formed:

$$RX^+ + RX \longrightarrow RXR^+ + X. \qquad (R41)$$

The thermochemistry of these reactions explains the different paths: the protonated parent ion will occur only if the X—H bond formed in RXH^+ is stronger than the C—H bond broken in RX. This is not the case for CH_3Br and CH_3I.

The monohalogen derivatives of ethane show the corresponding reactions. Mixed dialkylhalonium ions are also formed. As it possible that the halogen contains more than one isotope [29], double-resonance experiments indicate that the chlorine and bromine in the ionic product of the reactions (R38), (R39) and (R41) come with equal probability from the ionic and neutral reactants.

Reaction (R39) is a further example of a nucleophilic displacement reaction:

$$\left.\begin{array}{l} C_2H_5ClH^+ + CH_3Cl \\ \\ C_2H_5Cl + CH_3ClH^+ \end{array}\right\} \rightarrow \left[\begin{array}{l} CH_3CH_2\text{-}Cl \\ \vdots \\ CH_3\text{-}Cl\text{-}H^+ \end{array}\right] \rightarrow C_2H_5ClCH_3 \overset{+}{} + HCl. \qquad (R42)$$

The fluoromethanes show with exception of CF_4 the reactions (R38) and (R39). The dialkylfluoronium ion is the more destabilized the more hydrogen atoms are replaced through fluoride atoms. The fluoride transfer reactions can be used to determine carbonium ion stabilities $CHF_2^+ < CH_2F^+ < CF_3^+ < CH_3^+$. In mixtures of CH_3F and CH_2F_2 at low pressures three halonium ions $(CH_3)_2F^+$, $(CH_2F)_2F^+$ and $(CHF_2)_2F^+$ are observed. With increased pressure only $(CH_3)_2F^+$ remains:

$$(CH_2F)_2F^+ + CH_3F \longrightarrow CH_3FCH_2F^+ + CH_2F_2 \qquad (R43)$$

$$CH_3FCH_2F^+ + CH_3F \longrightarrow (CH_3)_2F^+ \quad + CH_2F_2. \qquad (R44)$$

In a mixture of CH_2F_2 and CHF_3 a reversible ion–molecule reaction occurs:

$$CHF_2^+ + CH_2F_2 \longrightarrow CH_2F^+ + CHF_3. \qquad (R45)$$

With a time-lag ion-ejection technique the rate constants for the forward (k_f) and the reverse (k_r) reaction have been determined:

$$k_f = 1.4 \cdot 10^{-10} \text{ cm}^3 \text{ molecule}^{-1} \text{ sec}^{-1}$$

$$k_r = 2.2 \cdot 10^{-10} \text{ cm}^3 \text{ molecule}^{-1} \text{ sec}^{-1}.$$

From the equilibrium constant $k = 0.64$ $\Delta G_{298}^0 = 0{,}25$ kcal mole^{-1} can be obtained.

Chloride and fluoride ion-transfer reactions in halomethanes and haloethanes have also been studied by Dawson, Henderson, O'Malley, and Jennings [247].

The most usual ion–molecule reaction observed in chloroethylene [29] is

$$A^+ + C_2H_3Cl \longrightarrow AC_2H_2^+ + HCl. \qquad (R46)$$

The reaction of $C_2H_3Cl^+$ with C_2H_3Cl produces: $C_3H_4Cl^+$, $C_4H_5Cl^+$ and $C_4H_6Cl^+$. The chlorine in the product comes with equal probability from the charged and the neutral reactant.

Vinyl fluoride [81] shows the corresponding ion–molecule reactions and loss of fluoromethyl radicals. From the parent ion, $C_3F_2H_3^+$ is formed in greatest abundance.

The ion–molecule reactions of C_2HF_3 and C_2F_4 [244] and of 1,1-difluoroethylene [245] and of ethylene with fluoroethylenes [242] have recently been studied.

The parent ions $C_2HF_3^+$ and CF_4^+ show only one condensation reaction, involving CF_3 elimination.

$$C_2HF_3^+ + C_2HF_3^+ \longrightarrow C_3H_2F_3^+ + CF_3 \qquad (R\,47)$$

$$C_2F_4^+ + C_2F_4^+ \longrightarrow C_3F_5^+ + CF_3 \qquad (R\,48)$$

The reaction of C_2F_4 can be interpreted also as the sum of a second- and a third-order process:

$$C_2F_4^+ + 2C_2F_4 \longrightarrow C_3F_5^+ + CF_3 + C_2F_4. \qquad (R\,49)$$

The molecular ions initially produced by electron impact contain excitation energy and are partially deexcited by non-reactive collisions and charge transfer before reacting.

In mixtures of C_2F_4 and C_2HF_3 a reversible charge-transfer reaction has been observed:

$$C_2F_4^+ + C_2HF_3 \rightleftharpoons C_2F_4 + C_2HF_3^+. \qquad (R\,50)$$

The free energy $\Delta G = 0.023 \pm 0.001$ eV was obtained by the method of Bowers *et al.*[147a].

Only one ion–molecule reaction of the molecular ion of 1,1-difluoroethylene can be observed [245]: formation of a collision-stabilized dimer ion, a third-order process with a very high rate constant which may arise from the high polarity of the molecule:

$$\begin{bmatrix} \overset{\delta+}{CH_2} = \overset{\delta-}{CF_2} \\ \vdots \quad\quad \vdots \\ \underset{CF_2}{\delta-} = \underset{CH_2}{\delta+} \end{bmatrix}^+$$

The molecular ions in mixtures of C_2H_4 with CH_2CHF, CH_2CF_2, $CHFCHF$, $CHFCF_2$ and C_2F_4 [252] react according to

$$M^+ + C_2H_4 \longrightarrow C_3(H,F)_5^+ + C(H,F)_3 \qquad (R\,51)$$

with hydrogen randomization. Introduction of fluorine leads in general to a fall in the rate constant of these reactions.

Reactions of the type

$$C_2H_4^+ + C_2F_4 \longrightarrow CH_2CF_2^+ + CH_2CF_2 \qquad (R\,52)$$

which proceed via a four-center complex show no randomization.

H. Aliphatic Alcohols

Methanol [47], ethanol, 2-propanol [91,173], 2-butanol [194] and tert-butanol [57] have been studied.

The main reactions starting from the protonated parent molecule or from a fragment ion resulting from α-cleavage, the main fragmentation process during ionization, are:

    a) condensation,

    b) polymerization to proton-bound dimers and trimers,

    c) dehydration by ions possessing a labile proton; this is an acid-catalyzed elimination process which does not occur in the case of ethanol.

Formation of a strong hydrogen bridge is a dominant process in all classes of reactions.

The 2-propanol reactions will be discussed in some detail.

The α-cleavage product, protonated acetaldehyde, reacts in two steps with i-propanol, and two water molecules are formed [173]:

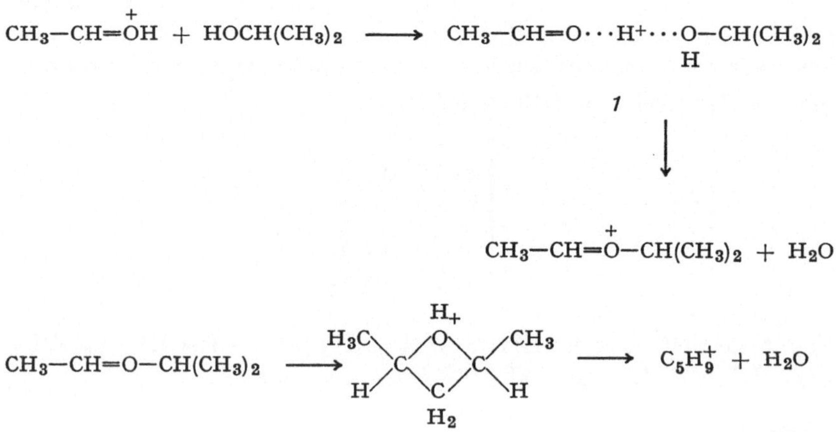

$$CH_3-CH{=}\overset{+}{O}H + HOCH(CH_3)_2 \longrightarrow CH_3-CH{=}O\cdots H^+\cdots \underset{H}{O}-CH(CH_3)_2$$

1

$$\downarrow$$

$$CH_3-CH{=}\overset{+}{O}-CH(CH_3)_2 + H_2O$$

$$CH_3-CH{=}O-CH(CH_3)_2 \longrightarrow \quad C_5H_9^+ + H_2O$$

$$(R\,53)$$

Proton-bound dimers and trimers are formed from the protonated alcohols and from other proton-bound compounds like:

$$(CH_3)_2CH\overset{+}{O}H_2 + (CH_3)_2CHOH \longrightarrow [(CH_3)_2CHOH]_2H^+$$

$$\Big\uparrow -CH_3CHO \qquad\qquad (R54)$$

$$1 + (CH_3)_2CHOH$$

The dehydration reaction leads to propene:

$$(CH_3)_2CH\overset{+}{O}H_2 + (CH_3)_2CHOH \longrightarrow (CH_3)_2CH\overset{+}{O}H_2\cdots OH_2$$

$$+ H_2C=CH-CH_3 \qquad (R55)$$

A similar condensation reaction occurs in a mixture of ethanol with butane-2,3-dione [132]. Not only the protonated ethanol but also the protonated diketone reacts with the corresponding neutral to form:

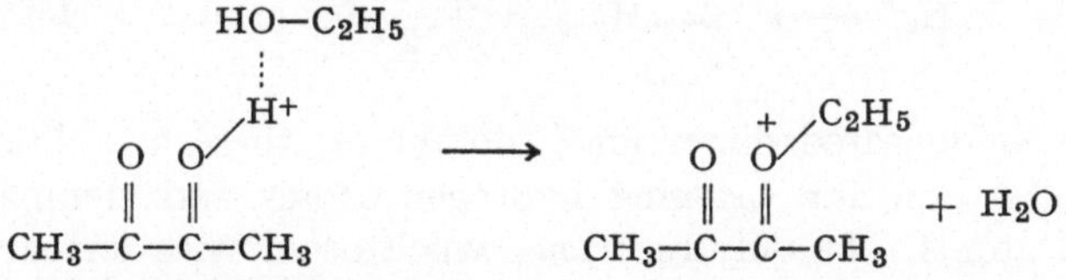

(R56)

The ion–molecule reactions of acetone [220] also produce protonated oligomers, which condense with the loss of 8 water molecules.

The protonated acetonitrile CH_3CNH^+ reacts with the neutral to form protonated dimer [46].

I. Inorganic Compounds

Hydrides

The gas-phase ion chemistry of several main-group hydrides has been studied:

Boranes [44,152,203]; silane [213,236,237]; ammonia [170,257]; phosphine [111,123]; arsine [264]; H_2S, H_2O [39], and H_2Se [201].

Dunbar [44,152,203] has studied the ion–molecule reactions of several boron hydrides: B_2H_6, B_4H_{10}, B_5H_9, B_5H_{11}, and B_6H_{10} and the gas-phase reactions of diborane with the same higher boron hydrides and with oxygen-containing compounds. The numerous reactions of the hydrides are readily classified into a few schemes.

The formation of negative ions as a function of electron energy shows for all compounds a sharp minimum at 5 eV. The negative product ions of diborane form a well-defined series: $B_2H_7^-$, $B_3H_8^-$, $B_4H_9^-$, $B_5H_{10}^-$, and $B_6H_9^-$, but no such regularity has been found for the higher boranes.

The negative ion–molecule reactions strongly support the classification of the boron hydrides into a B_nH_{n+4} (stable hydrides) and a B_nH_{n+6} (unstable hydrides) series. While the first-named form $(M-1)^-$ ions preferentially, the latter form a very stable molecular ion and by symmetrical cleavage $(M-BH_3)^-$ ions and product ions with successive addition of BH groups, e.g.

$$B_5H_{\overline{11}} + B_5H_{11} \longrightarrow B_6H_{\overline{12}} + B_4H_{10}. \tag{R57}$$

Ion–molecule reactions of the cations lead, in contrast to hydrocarbon chemistry, to highly electron-deficient odd-electron products, mainly by a condensation reaction with H elimination:

$$B_xH_y^+ + B_uH_v \longrightarrow B_{x+u}H_{y+v-n}^+ + \frac{n}{2}H_2 \quad (n = 1\ldots.10) \tag{R58}$$

The "fully condensated dimer ions" dominate; these are obtained by retaining all boron and bridging hydrogen atoms and eliminating all terminal H atoms. The only reactions which occur with formation of a neutral boron-containing product (BH_3) leading to borane ions $B_3H_n^+$ ($n = 3,4,5,6$) are endothermic and appear in the case of diborane. They can only be detected by double-resonance experiments if the primary ions have sufficient internal energy.

In a mixture of H_2O and B_2H_6 the only anion formed in great abundance is $B_2H_5^-$. Dunbar [203] suggested the following mechanism:

$$OH^- + B_2H_6 \longrightarrow (BH_4^-)^* + \text{neutrals} \tag{R59}$$

$$(BH_4^-)^* + B_2H_6 \begin{cases} \nearrow B_2H_5^- \\ \searrow B_2H_7^- \end{cases} \tag{R60}$$

All the ions formed by electron impact on SiH_4 react with the neutral yielding in most cases addition products [213] and hydrogen molecules:

$$SiH_n^+ + SiH_4 \longrightarrow SiH_{4+n-m} + \frac{m}{2}H_2. \tag{R61}$$

The reactivity of Si^+ is remarkable:

$$Si^+ + SiH_4 \longrightarrow Si_2H_2^+ + H_2 \tag{R 62}$$

The rate constant of the process:

$$SiH_3^+ + SiH_4 \longrightarrow SiH_4 + SiH_3^+ \tag{R 63}$$

is large. The H^- transfer is a far more important mechanism than in the case of the corresponding hydrocarbons, where proton transfer is observed.

No SiH_5^+ is formed in the pure silane, perhaps due to the low abundance of SiH_4^+, but may be formed in mixtures with CH_4[236]. In the last-named mixtures, and in mixtures with CD_4[237], two main mechanisms occur: H^- transfer reactions to CH_4^+, CH_4^+ and CH_5^+ from SiH_4 to produce SiH_3^+ with large rate constants and formation of methylsilanes like:

$$CH_4^+ + SiH_4 \longrightarrow SiCH_5^+ + H_2 + H. \tag{R 64}$$

In NH_3 mainly four ion–molecule reactions are found [170,247].

$$NH_3^+ + NH_3 \begin{array}{l} \overset{k_1}{\nearrow} NH_4^+ + NH_2 \qquad (R\,65\,a) \\ \underset{k_2}{\searrow} NH_3^+ + NH_3 \qquad (R\,65\,b) \end{array}$$

$$NH_2^+ + NH_3 \begin{array}{l} \overset{k_3}{\nearrow} NH_4^+ + NH \qquad (R\,66\,a) \\ \underset{k_4}{\searrow} NH_3^+ + NH_2 \qquad (R\,66\,b) \end{array}$$

With low-abundance condensation products, $N_2H_n^+$ ($n = 1..5$) occur [123]. The charge-transfer reaction (R 65 b) takes place only for translationally excited ions [170]. k_1 and k_3 decrease and, as expected, k_2 and k_3 increase with increasing primary-ion kinetic energy [170]. Marx and Mauclaire [257] have measured a rate constant for the reaction which is larger by a factor of two than the result of Huntress et al.[170]. They proposed that the production of NH_3^+ ions during NH_2^+ ejection must be taken in account. No ion–molecule reactions of negative ions of NH_3 have been observed [247].

PH_3 [111,123] shows two types of ion–molecule reactions: proton transfer of the type

$$PH_3^+ + PH_3 \longrightarrow PH_4^+ + PH_2 \tag{R 67}$$

forming from PH_3 solely PH_4^+, which does not react further, and condensation reactions very like those observed for CH_4: $P_2H_n^+$ ($n = 0...5$) and P_3H_m ($m = 0..2$).

At low pressures only the secondary ion PH_4^+ is observed; when the pressure is increased the abundance of the other secondary ions goes through a maximum. At pressures of 10^{-3} torr all secondary ions react to give PH_4^+.

In mixtures with H_2O, NH_3 and CH_4 the main condensation products are: POH^+, POH_2^+, PNH_2^+, PNH_3^+, PCH_3^+ and PCH_4^+, appearing from reaction with PH^+.

The ion chemistry of AsH_3 [264] and PH_3 shows great similarity with H transfer to form AsH_n^+ ($n = 0...5$) and condensation reactions $As_2H_n^+$ ($n = 0..5$) and As_3^+ and As_3H^+. At higher pressures, however, the condensation products become dominant. The negative ion chemistry of arsine shows only two ion–molecule reactions:

$$As^- + AsH_3 \longrightarrow As_2H^- + H_2 \qquad (R\,68)$$

$$AsH^- + AsH_3 \longrightarrow AsH_2^- + H_2 \qquad (R\,69)$$

AsH_2^- is unreactive towards AsH_3.

As already stated for the group-iv hydrides, CH_4 and SiH_4, there is a pronounced difference between the ion chemistry of the first group-v hydride and the following PH_3 and AsH_3. However, the ion chemistry of PH_3, AsH_3 and CH_4 is very similar.

NH_3 undergoes mainly charge transfer and proton transfer; NH_4^+ is formed by fragment ions, too. On the other hand, PH_4^+ and AsH_4^+ are formed only from the corresponding parent ions. The condensation products of NH_3 are in very low abundance. Only $N_2H_x^+$ are formed. PH_3, on the contrary, condenses to di-and triphosphines in abundance, passing through a maximum at intermediate pressures. Finally, condensation is the most abundant ion–molecule reaction for AsH_3, the higher condensated products becoming more abundant with increasing pressure.

Very few ion–molecule reactions have been observed for H_2O [39], H_2S [39] and H_2Se [201]. They are listed in Table 6.

As with group-v hydrides, the expulsion of H_2 instead of H becomes thermodynamically more favorable with increasing atomic number. In contrast to OH^+ and SH^+, SeH^+ does not react with the corresponding neutral. No H_3Se^+ is formed, even though this reaction is exothermic.

Other Inorganic Compounds

The principal ion–molecule reaction in HCN [74] is the protonation of the molecular ion:

$$HCN^+ + HCN \longrightarrow H_2CN^+ + CN. \qquad (R\,70)$$

The O^- ion, formed by dissociative attachment from N_2O [258], shows only one ion–molecule reaction:

$$O^- + NNO \longrightarrow NO^- + NO \tag{R71}$$

and a collision detachment at higher pressures:

$$NO^- + NNO \longrightarrow NO + NNO + e. \tag{R72}$$

The rate constant k shows a low- and a high-pressure limit, because O^- is initially formed with excess energy. A simple kinetic treatment has been used to calculate the rate constant and its energy dependence, in good agreement with experimental data.

The sole study on transition metal carbonyl ion chemistry with ICR mass spectrometry has been published by Forster and Beauchamp [157]. The ions produced by electron impact: Fe^+, $Fe(CO)^+$ and $Fe(CO)_2^+$ in pure $Fe(CO)_5$, undergo two polymerization reactions to yield $Fe(CO)_4^+$ and $Fe_2(CO)_5^+$. In binary mixtures with CH_3F, H_2O and NH_3, substitution reactions occur like:

$$Fe(CO)_n^+ + CH_3F \longrightarrow Fe(CH_3F)(CO)_{n-1}^+ + CO\,(n = 1\ldots4) \tag{R73}$$

and during an oxidative addition reaction:

$$(CH_3)_2F^+ + Fe(CO)_5^+ \longrightarrow Fe(CH_3)(CO)_5^+ + CH_3 \tag{R74}$$

the formal oxidation state of the iron atom increases by two. The species $HFe(CO)_5^+$ and $HFe(CO)_4^+$ are observed, derived from H_3O^+ by proton exchange. No reaction with HCl takes place. This fact is rationalized by correlating the proton affinities of the ligands studied with their ability to effect substitution.

The ion chemistry of thiothionylfluoride [188] is characterized by sulfur transfer reactions.

$$S_2F_2^+ + SSF_2 \longrightarrow S_3F_2^+ + SF_2 \tag{R75a}$$
$$\longrightarrow S_3F^+ + SF_3 \tag{R75b}$$
$$\longrightarrow S_2F_3^+ + S_2F \tag{R75c}$$
$$S_3F_2^+ + SSF_2 \longrightarrow S_4F_2^+ + SF_2 \tag{R76}$$

The positive and negative ions formed from sulfur tetrafluoride [243] react mainly via the F-transfer mechanism. An ICR spectrum of the positive and negative ions from SF_4 is shown by way of example in Fig. 2. The reaction

$$SF_4^- + SF_4 \longrightarrow SF_5^- + SF_3 \qquad (R77)$$

is the reaction with greatest abundance. At relatively high pressures a tertiary ion, SF_6^-, is also observed

$$SF_5^- + SF_4 \longrightarrow SF_6^- + SF_3. \qquad (R78)$$

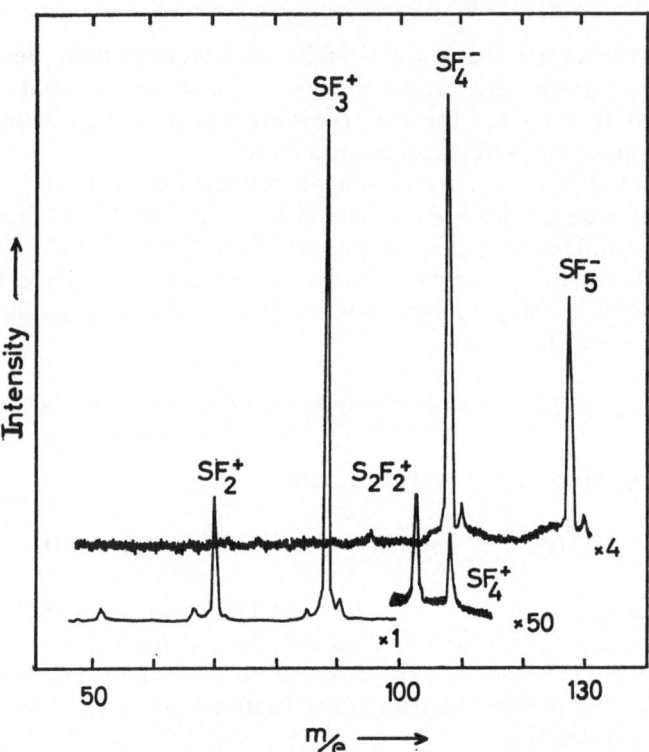

Fig. 2. Positive and negative ions in SF_4

As can be seen in Fig. 3, the pressure-broadening of the ICR lines is marked. The corresponding ion cyclotron double-resonance spectrum in Fig. 4 depicts the changes in SF_6^- signal intensity while sweeping the radiofrequency of the rf transmitter over the range of SF_4^- and SF_5^- resonance frequencies.

Acknowledgement. The authors express their gratitude to Prof. J. L. Beauchamp, Prof. J. H. Futrell, Prof. K. R. Jennings, Prof. R. Marx and Prof. S. Wexler for preprints of their unpublished results.

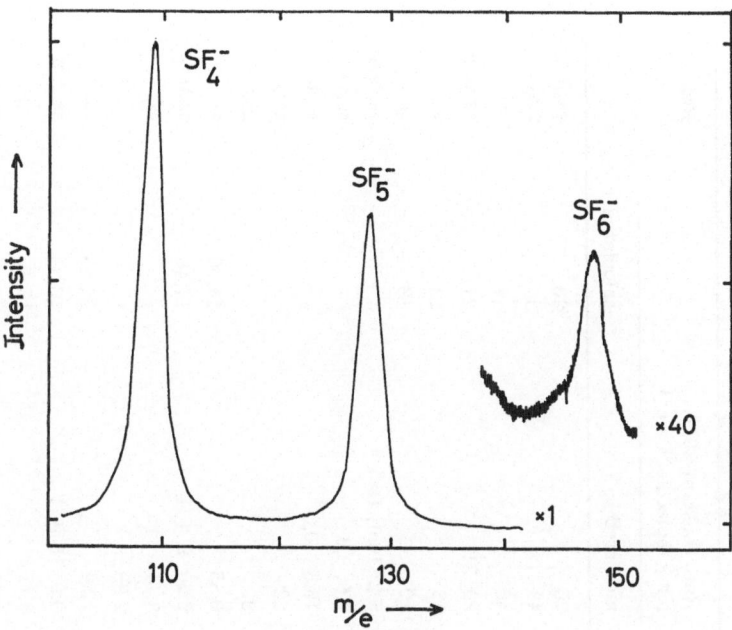

Fig. 3. Negative ions in SF_4 at higher pressure

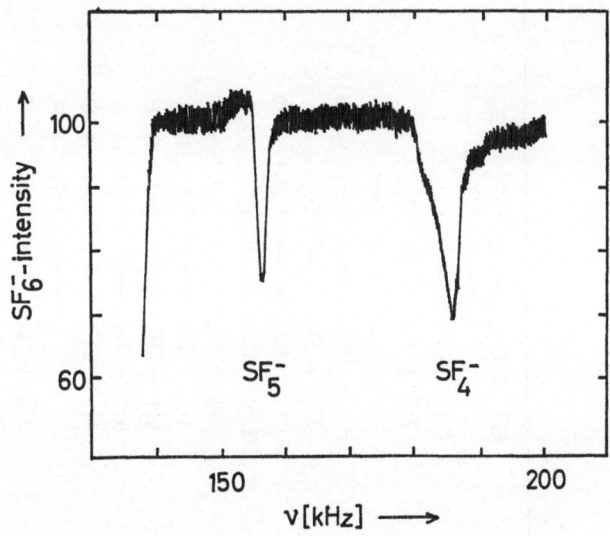

Fig. 4. Double-resonance observation of SF_6^- in SF_4 at higher pressure

99

Table 5. Thermal energy rate constants determined with ICR spectroscopy

Reaction	Rate constant 10^{10} (cm^3 molecule^{-1} sec^{-1})		Ref.
	measured	calculated[1]	
$H_2^+ + CO_2 \longrightarrow CO_2^+ + H_2$	20	36	262)
$\longrightarrow CO_2H^+ + H$	49	36	262)
$CO_2^+ + H_2 \longrightarrow CO_2H^+ + H$	40	14	262)
$H_2^+ + H_2 \longrightarrow H_3^+ + H$	20, 21	21	62,262)
$O^+ + CO_2 \longrightarrow$ prod	15	12	262)
$CH_4^+ + CH_4 \longrightarrow CH_5^+ + CH_3$	10	13	262)
$D_2^+ + D_2 \longrightarrow D_3^+ + D$	5.2; 16,0; 16		76,62,200)
$HD^+ + HD \longrightarrow H_2D^+ + D$	7.5; 8,0		62,200)
$HD^+ + HD \longrightarrow HD_2^+ + H$	10.5; 10,0		62,200)
$H_2^+ + D_2 \longrightarrow H_2D^+ + D$	32		200)
$D_2^+ + H_2 \longrightarrow HD_2^+ + H$	30		200)
$Ar^+ + H_2 \longrightarrow ArH^+ + H$	6.85	15.60	63)
$Ar^+ + D_2 \longrightarrow ArD^+ + D$	6.20; 9 ± 1	11.40	63,107)
$Ar + HD \longrightarrow ArH^+ + D$	3.14		63)
$\longrightarrow ArD^+ + H$	2.98		63)
$H_2^+ + Ar \longrightarrow ArH^+ + H$	12.4	22.50	63)
$D_2^+ + Ar \longrightarrow ArD^+ + D$	9.9; 16 ± 1	16.3	63,107)
$HD^+ + Ar \longrightarrow ArH^+ + D$	6.11	18.6	63)
$\longrightarrow ArD^+ + H$	6.28	18.6	63)
$H_3^+ + Ar \longrightarrow ArH^+ + H_2$	3.65	19.1	63)

Reaction			Ref.
$D_3^+ + Ar \rightarrow ArD^+ + D_2$	4.71	13.5	(63)
$N_2^+ + H_2 \rightarrow N_2H^+ + H$	14.10	15.40	(63)
$N_2^+ + D_2 \rightarrow N_2D^+ + D$	12.60	11.20	(63)
$N_2^+ + HD \rightarrow N_2H^+ + D$	5.63	13.5	(63)
$\rightarrow N_2D^+ + H$	5.46	13.5	(63)
$H_2^+ + N_2 \rightarrow N_2H^+ + H$	19.5	22.80	(63)
$D_2^+ + N_2 \rightarrow N_2D^+ + D$	16.10	16.60	(63)
$HD^+ + N_2 \rightarrow N_2H^+ + D$	8.05	20.00	(63)
$\rightarrow N_2D^+ + H$	8.24	20.00	(63)
$H_3^+ + N_2 \rightarrow N_2H^+ + H_2$	10.30	19.00	(63)
$D_3^+ + N_2 \rightarrow N_2D^+ + D_2$	7.49	13.95	(63)
$N_2O^+ + H_2 \rightarrow N_2OH^+ + H$	3.4 ± 0.7	12.9	(259)
$H_2^+ + N_2O \rightarrow N_2OH^+ + H$	7.9 ± 1.6	28.7	(259)
$N_2O^+ + CH_4 \rightarrow N_2OH^+ + CH_3$	9.5 ± 1.9	10.5	(259)
$CH_4^+ + N_2O \rightarrow N_2OH^+ + CH_3$	10.1 ± 2	11.6	(259)
$CH_4^+ + N_2O \rightarrow NOH^+ + CH_3N$	3.0 ± 0.15		(259)
$N_2O^+ + CH_4 \rightarrow NOH^+ + CH_3N$	3.0 ± 0.15		(259)
$N_2O^+ + H_2 \rightarrow N_2H^+ + OH$	1.4		(259)
$H_2^+ + N_2O \rightarrow N_2H^+ + OH$	4.7		(259)
$CH_4^+ + CH_4 \rightarrow CH_5^+ + CH_3$	12 ± 1	13.2	(107,182)
$CH_3^+ + CH_4 \rightarrow C_2H_5^+ + H_2$	$10 \pm 1; 13$	13.5	(107)
$CH_4^+ + D_2 \rightarrow CH_4D^+ + D$	9.0		(76)
$CH_2D_2^+ + CH_2D_2 \rightarrow CH_3D_2^+ + CHD_2$	5.1		(67)
$CH_2D_2^+ + CH_2D_2 \rightarrow CH_2D_3^+ + CH_2D$	4.1		(67)
$C_2H_2^+ + C_2H_2 \rightarrow C_4H_2^+ + H_2$	3.9		(67)
$C_2H_2^+ + C_2H_2 \rightarrow C_4H_3^+ + H$	8.3		(67)

Table 5 (continued)

Reaction	Rate constant 10^{10} (cm^3 molecule^{-1} sec^{-1}) measured	calculated[1]	Ref.
$CD_4^+ + CD_4 \longrightarrow CD_5^+ + CD_3$	11	12.1	253)
$CD_3^+ + CD_4 \longrightarrow C_2D_5^+ + D_2$	12	12.4	253)
$CH_3^+ + NH_3 \longrightarrow CH_4N^+ + H_2$	20	12.6	125)
$CH_3FH^+ + HCl \longrightarrow CH_3ClH^+ + HF$	3.1		124)
$+ N_2 \longrightarrow CH_3N_2^+ + FH$	14		168)
$+ CH_3F \longrightarrow (CH_3)_2F^+ + HF$	7.1		168)
$+ CO \longrightarrow CH_3CO^+ + HF$	0.56		168)
$CH_3^+ + CH_3COCH_3 \longrightarrow prod$	20	16.8	220)
$CH_3CO^+ + CH_3COCH_3 \longrightarrow prod$	4.3	11.8	220)
$CH_3COCH_3^+ + CH_3COCH_3 \longrightarrow prod$	5.4	10.9	220)
$CH_3N_2^+ + CH_3N_2CH_3 \longrightarrow prod$	5.1		206)
$CH_3N_2CH_3^+ + CH_3N_2CH_3 \longrightarrow (CH_3) N=NCH_3^+ + CH_3 + N_2$	0.038		206)
$CH_3F^+ + CH_3F \longrightarrow CH_3FH^+ + CH_3F$	13.6; 12.8	9.07	137,193)
$\longrightarrow$		17.3	246)
$\longrightarrow C_2H_4F^+ + HF + H$	0.96; 0.96	17.3	246)
$\longrightarrow C_2H_6F^+ + F$	8.0; 7.1	9.0	137)
$CH_3FH^+ \longrightarrow prod$	9.9		246)
$C_2H_5F^+ + C_2H_5F \longrightarrow C_2H_5FH^+ + C_2H_4F$	15	10.0	193)
$C_2H_5FH^+ + C_2H_5F \longrightarrow C_4H_{10}F^+ + HF$	14	9.94	193)
$CH_3Cl^+ + CH_3Cl \longrightarrow CH_3ClH^+ + CH_2Cl$	12.5	9.79	193)

Reaction				Ref.
$CH_3ClH^+ + CH_3Cl$	$\rightarrow C_2H_6Cl^+ + HCl$	1.4	9.74	193)
$C_2H_5Cl^+ + C_2H_5Cl$	$\rightarrow$ prod	14.95	10.32	193)
$C_2H_5ClH^+ + C_2H_5Cl$	$\rightarrow C_4H_{10}Cl^+ + HCl$	5.5	10.28	193)
$CH_3Br^+ + CH_3Br$	$\rightarrow C_2H_6Br^+ + Br$	0.55	7.91	193)
$C_2H_5Br^+ + C_2H_5Br$	$\rightarrow C_2H_{10}Br^+ + Br$	0.56	8.56	193)
$CH_3I^+ + CH_3I$	$\rightarrow C_2H_6I^+ + I$	0.055	7.98	193)
$C_2H_5I^+ + C_2H_5I$	$\rightarrow C_4H_{10}I^+ + I$	0.16	7.98	193)
$CHF_2^+ + CH_2F_2$	$\rightarrow CH_2F^+ + CF_3H$	1.9		246)
$CH_2F_2^+ + CH_2F_2$	$\rightarrow CH_2F_2H^+ + CF_2H$	13.0		246)
$CH_2F_2H^+ + CH_2F_2$	$\rightarrow (CH_2F)_2F^+ + F^+ + HF$	14.0		246)
$CF_3^+ + CHF_3$	$\rightarrow CHF_2^+ + CF_4$	2.1		246)
$CH_2F_3H^+ + CF_3H$	$\rightarrow (CHF_2)F^+ + HF$	0.82		246)
$CF_2^+ + CF_4$	$\rightarrow CF_3^+ + CF_3$	1.7		246)
$C_2HF_3^+ + C_2HF_3$	$\rightarrow C_3H_2F_3^+ + CF_3$	0.56		245)
$C_3H_2F_3^+ + C_2HF_3$	$\rightarrow C_3HF_3^+ + C_2H_2F_2$	0.30		245)
$C_2F_4^+ + C_2F_4$	$\rightarrow C_3F_5^+ + CF_3$	0.10		235)
$C_2F_4^+ + C_2HF_3$	$\rightarrow C_2HF_3^+ + C_2F_4$	0.99		245)
$C_2HF_3^+ + C_2F_4$	$\rightarrow C_2F_4^+ + C_2HF_2$	0.52		245)
$C_2F_4^+ + C_2HF_3$	$\rightarrow C_3HF_4^+ + CF_3$	0.07		245)
$C_2HF_3^+ + C_2F_4$	$\rightarrow C_3HF_4^+ + CF_3$	0.07		245)
$C_2F_2H_2^+ + C_2H_2F_2$	$\rightarrow C_4H_4F_4^+$	3.1 ± 0.4 10^{-24}, 2)		252)
$C_2H_4^+ + C_2H_4$	$\rightarrow C_3H_5^+ + CH_3$	8.0		252)
$CH_2CHF^+ + C_2H_4$	$\rightarrow$ prod	5.9 ± 1.8		252)
$CH_2CF_2^+ + C_2H_4$	$\rightarrow$ prod	2.5 ± 0.5		252)
$CHF CHF^+ + C_2H_4$	$\rightarrow$ prod	2.4 ± 0.5		252)
$CHFCF_2^+ + C_2H_4$	$\rightarrow$ prod	6.7 ± 0.7		252)

103

Table 5 (continued)

Reaction	Rate constant 10^{10} (cm³ molecule⁻¹ sec⁻¹)		Ref.
	measured	calculated[1]	
$C_2F_4^+ + C_2H_4 \longrightarrow$ prod	1.07 ± 0.3		252)
$SiH_2^+ + SiH_4 \longrightarrow SiH_3^+ + SiH_3$	10.7 ± 2.0		213)
$\longrightarrow Si_2H_2^+ + 2H_2$	2.1 ± 0.3		213)
$\longrightarrow Si_2H_4^+ + H_2$	2.5		213)
$Si^+ + SiH_4 \longrightarrow Si_2H_2^+ + H_2$	4.8 ± 0.6		213)
$SiH^+ + SiH_4 \longrightarrow Si_2H_3^+ + H_2$	2.8 ± 0.07		213)
$SiH_3^+ + SiH_4 \longrightarrow Si_2H_5^+ + H_2$	0.24 ± 0.04		213)
$SiH_3^+ + SiH_4 \longrightarrow Si_2H_3^+ + 2H_2$	0.07 ± 0.04		213)
$SiH^+ + SiH_4 \longrightarrow Si_2H^+ + 2H_2$	0.7 ± 0.2		213)
$SiH_3^+ + SiH_4 \longrightarrow SiH_3^+ + SiH_4$	13.5 ± 0.5		213)
$CH^+ + SiH_4 \longrightarrow SiCH_2^+ + H_2 + H$	1.5 ± 0.4		236)
$Si^+ + CH_4 \longrightarrow SiCH_3^+ + H_2$	0.82 ± 0.28		236)
$CH_2^+ + SiH_4 \longrightarrow SiCH_3^+ + H_2 + H$	2.91 ± 0.53		236)
$SiH_2^+ + CH_4 \longrightarrow SiCH_4^+ + H_2$	3.07 ± 0.61		236)
$\longrightarrow SiCH_5^+ + H$	0.52 ± 0.15		236)
$CH^+ + SiH_4 \longrightarrow SiH_3^+ + CH_4$	11.07 ± 1.8		236)
$CH_4^+ + SiH_4 \longrightarrow SiCH_4^+ + CH_2$	0.35 ± 0.23		236)
$\longrightarrow SiCH_5^+ + H_2 + H$	0.30 ± 0.18		236)
$\longrightarrow SiH_3^+ + CH_3 + H_2$	25 ± 10		236)
$CH_5^+ + SiH_4 \longrightarrow SiH_3^+ + CH_4 + H_2$	17.8 ± 2.5		236)

$NH_3^+ + NH_3$	$\rightarrow$ $NH_4^+ + NH_2$	$1.9 \pm 0.2; 14 \pm 15$	170,257)
	$\rightarrow$ $NH_3^+ + NH_3$	0.4	170)
$NH_2^+ + NH_3$	$\rightarrow$ $NH_4^+ + NH$	$1.1 \pm 0.2; 9 \pm 1.5$	170,257)
	$\rightarrow$ $NH_3^+ + NH_2$	$1.1 \pm 0.2; 21 \pm 4$	170,257)
$P^+ + PH_3$	$\rightarrow$ $P_2H^+ + H_2$	8.2	123)
$PH^+ + PH_3$	$\rightarrow$ $P_2^+ + 2H_2$	4.0	123)
	$\rightarrow$ $P_2H_2^+ + H_2$	3.2	123)
	$\rightarrow$ $P_2H_2^+ + H$	1.8	123)
$PH_2^+ + PH_3$	$\rightarrow$ $P_2H^+ + 2H_2$	6.2	123)
	$\rightarrow$ $P_2H_3^+ + H_2$	2.2	123)
$PH_3^+ + PH_3$	$\rightarrow$ $PH_4^+ + PH_2$	10.5	123)
$H_2Se^+ + H_2Se$	$\rightarrow$ $H_3Se^+ + HSe$	4 ± 1	201)
$HSe^+ + H_2Se$	$\rightarrow$ $HSe_2^+ + H_2$	3 ± 1	201)
$Se^+ + H_2Se$	$\rightarrow$ $Se_2^+ + H_2$	4 ± 1	201)
$HCN^+ + HCN$	$\rightarrow$ $H_2CN^+ + CN$	12	74)
$C^+ + HCN$	$\rightarrow$ $C_2N^+ + H$	13	74)
$O^- + N_2O$	$\rightarrow$ $NO^- + NO$	3.0	258)
$NO^- + N_2O$	$\rightarrow$ $NO + N_2O + e^-$	0.051	258)
$S_2F_2^+ + SSF_2$	$\rightarrow$ $S_3F_2^+ + SF_2$	20	188)
$CH_3FH^+ + HCl$	$\rightarrow$ $CH_3ClH^+ + HF$	3.1 ± 0.3	124)

1) Calculated with $k = 2\pi e \left(\dfrac{\mu}{\alpha}\right)^{\frac{1}{2}}$.

2) Third-order rate constant.

Table 6. Molecules studied with ICR spectroscopy

Compound	Ref.
H_2	34,62,63,64,76,94,96,102,107, 147b,159,171,200,259,262)
D_2	62,63,76,95,96,262)
HD	62,63,200)
Rare gases	50,63,68,95,107,167)
N_2	50,63,64,79,82,155,171)
CO_2	102,140,171,262)
N_2O	153,258,259)
Boron hydrides	44,152,203)
SiH_4	213,236,237)
NH_3	65,67,123,125,170,257,234)
NF_3	169)
PH_3	65,111,123,233)
AsH_3	234,264)
H_2S	39,106)
H_2O	39,42,140,149b)
H_2Se	201,245)
$Fe(CO)_5$	157)
SSF_2	188)
SF_6	79,243,120)
HCN	74)
Methane	55,61,67,68,76,94,107,125,130, 155,167,214,216,239,253)
Acetylene	67,80,96,106,149b)
Ethylene	40,58,81,106,114,115,118,119, 147b,205,215)
Ethane	189,204,214,224)
Allene	100)
Cyclopropane	181)
Propyne	100)
Butene	118,119,216)
Pentenes	118,119,164,208)
Hexenes	118,119)
Cyclooctatetraene	190)
Methylhalides	130,153,193,247,248)
Fluoromethanes	136,137,193,206,216,246,260)
Hexafluoroethane	51,187,216,245)
Ethylhalides	193)
Fluoroethylenes	24,81,244,252)
Chloroethylene	29,37,193)

Table 6 (continued)

Compound	Ref.
Octafluoropropane	187,216)
Perfluorocyclobutane	216)
Acetonitril	45,46)
Azomethane	206)
Lower aliphatic amines	65,192,212)
Ethylnitrate	92,128,148,149)
Pyridine	238)
Methanol	42,47,109,184)
Ethanol	42,149,184)
2-Propanol	91,173)
2-Butanol	194)
tert-Butanol	42,57)
1-Methylcyclobutanol	69,220)
2-Vinyl-1-methyl-cyclobutane-1-ol	219)
Ethylenoxide	96)
Acetone	69,220)
Acetaldehyde	147b)
2-Propylcyclopentanone	211)
Butanedion-2,3	122,131,156)
Pent-4-en-2-one	219)
2-Hexanone	69,220)
4-Nonanone	110)
Methyl ethyl ether	91)
Benzene and derivatives	82,85,150,190,248,250,149b)

Table 7. Binary mixtures studied with ICR spectrometry

Mixture	Ref.
H_2 and D_2	62,200)
$(H,D)_2$ and A	63,95)
H_2 and NNO	249)
H_2 and CO_2	251)
$(H,D)_2$ and $C(H,D)_4$	76,107)
$(H,D)_2$ and $C_2(H,D)_6$	189)
H_2 and C_2H_4O, CH_3CHO	147b)
B_2H_6 and CH_3OH, C_2H_5OH, H_2O	203)
SiH_4 and CH_4	236,237)

Table 7 (continued)

Mixture	Ref.
PH_3 and H_2O, NH_3, CH_4	123)
AsH_3 and PH_3, propene, CH_3Cl, H_2S	264)
H_2O and ethylene, ethanol, formaldehyde, ethylbromide	39)
H_2S and ethylene, acetylene	106)
$Fe(CO)_5$ and CH_3Cl, HCl, NH_3, H_2O	157)
Methane and Xe	68)
Methane and fluoromethanes	253,260)
Fluoromethanes	247)
Ethylene and acetylene	40)
Ethylene and fluoroethylenes	252)
Trifluoroethylene and tetrafluoroethylene	245)
Acetonitrile and methanol, acetaldehyde, diethyl ether	45)
Azomethane and NH_3, methylamine	206)
2-Propanol, acetone and ethylene oxide, acetaldehyde methyl ethyl ether	91)
2-Propylcyclopentanone and cyclohexanone, acetone	211)
Tetrahydrofurane and $N(H,D)_3$	207)
Furane and propylhalides	163)
Propylhalides and methanol, 2-propanol	221)
1,3-Butadiene and 1-pentene, 3-methyl-1-butene *cis*, *trans* 2-pentene	164,208)
Alkylnitrates and toluene, cycloheptratriene, norbornadiene	166)

References

1) Bleakney, W., Hipple, J. A.: Phys. Rev. *53*, 521 (1938).
2) Bloch, F.: Phys. Rev. *70*, 460 (1946).
3) Goudsmit, S. A.: Phys. Rev. *74*, 622 (1949).
4) Hipple, J. H., Thomas, H. A.: Phys. Rev. *75*, 1616 (1949).
5) Hipple, J. A., Sommer, H., Thomas, H. A.: Phys. Rev. *76*, 1877 (1949).
6) Monk, G. W., Werner, G. C.: Rev. Sci. Instr. *20*, 93 (1959).
7) Pound, R. V., Knight, W. D.: Rev. Sci. Instr. *21*, 219 (1950).
8) Smith, L. G.: Phys. Rev. *81*, 295 (1951).
9) Sommer, H., Thomas, H. A., Hipple, J. A.: Phys. Rev. *82*, 697 (1951).
9a) Berry, C. E.: J. Appl. Phys. *25*, 28 (1954).
10) Allis, W. P., (ed. S. Flügge): Handbuch der Physik XXI (1956).
11) Robinson, F., Hall, L. G.: Rev. Sci. Instr. *27*, 504 (1956).
12) McConnel, H. M.: J. Chem. Phys. *28*, 430 (1958).
13) Robinson, F. N. H.: J. Sci. Instr. *36*, 481 (1959).

14) Leont'ev, N. I.: Instr. Exptl. Tech. (USSR) (English Transl.) *5*, 788 (1961).
15) Bayes, K. D., Kivelson, D., Wong, S. C.: J. Chem. Phys. *37*, 1217 (1962).
16) Graham, J. R. Jr., Malone, D. P., Wobschall, D. C.: Bull. Am. Phys. Soc. *7*, 69 (1962).
17) Graham, J. R., Jr., Malone, D. P., Wobschall, D. C.: Bull. Am. Phys. Soc. *8*, 444 (1963).
18) Wobschall, D., Graham, J. R., Malone, D. P.: Phys. Rev. *131*, 1565 (1963).
19) McDaniel, E. W.: Collision phenomena in ionized gases. New York: Wiley 1964.
20) McDaniel, E. W.: Mass spectrometer uses new approach. Chem. Eng. News *43*, 21—55 (1965).
21) Wobschall, D.: Rev. Sci. Instr. *36*, 466 (1965).
22) Wobschall, D., Graham, J. R.: Bull. Am. Phys. Soc. *10*, 190 (1965).
23) Wobschall, D., Graham, J. R., Malone, D. P.: J. Chem. Phys. *42*, 3955 (1965).
24) Anders, L. R., Beauchamp, J. L., Dunbar, R. C., Baldeschwieler, J. D.: J. Chem. Phys. *45*, 1062 (1966).
25) Bibliography of atomic and molecular processes Jan—June 1967. ORNL-AMPIC-90 AK Ridge Nat. Lab. Tenn. 225 (1967).
26) Anders, L. R.: Thesis, Harvard Univ. Cambridge, Mass 1967.
27) Baldeschwieler, J.: Kagaku No Ryoiki *21*, 785 (1967).
28) Beauchamp, J. L.: J. Chem. Phys. *46*, 1231 (1967).
29) Beauchamp, J. L., Anders, L. R., Baldeschwieler, J. D.: J. Am. Chem. Soc. *89*, 4569 (1967).
30) Baldeschwieler, J. D.: 51-K-001-107 (1967).
31) —.
32) Dunbar, R. C.: J. Chem. Phys. *47*, 5445 (1967).
33) Flügge, R. A.: U.S. Govt. Res. Dev. Rept. *69*, 63 (1969).
34) Wobschall, D., Fluegge, R. A., Graham, J. R., Jr.: J. Chem. Phys. *47*, 4091 (1967).
35) Wobschall, D., Fluegge, R. A., Graham, J. R.: J. Appl. Phys. *38*, 3761 (1967).
36) Baldeschwieler, J. D.: Science *159*, 263 (1968).
37) Baldeschwieler, J. D., Benz, H., Llewellyn, P. M.: Advan. Mass Spectrometry *4*, 113 (1968).
38) Beauchamp, J. L., Noyes, A. A.: CALT-757-3 Calif. Inst. Tech., Calif USA (1968).
39) Beauchamp, J. L., Butrill, S. E.: J. Chem. Phys. *48*, 1783 (1968).
40) Bowers, M. T., Ellemann, D. D., Beauchamp, J. L.: J. Chem. Phys. *72*, 3599 (1968).
41) Bowers, M. T., Elleman, D. D., Miller, J. A.: Am. Chem. Soc. Nat. Meeting, April 1968, Sect. S.
42) Brauman, J. I., Blair, L. K.: J. Am. Chem. Soc. *90*, 6561 (1968).
43) Brauman, J. I., Blair, L. K.: *90*, 5636 (1968).
44) Dunbar, R. C.: J. Am. Chem. Soc. *90*, 5676 (1968).
45) Gray, G. A.: J. Am. Chem. Soc. *90*, 2177 (1968).
46) Gray, G. A.: J. Am. Chem. Soc. *90*, 6002 (1968).
47) Henis, J. M. S.: J. Am. Chem. Soc. *90*, 844 (1968).
48) Henis, J. M. S., Frasure, W.: Rev. Sci. Instr. *39*, 1772 (1968).
49) Kaplan, F.: J. Am. Chem. Soc. *90*, 4483 (1968).
50) King, J., Elleman, D. D.: J. Chem. Phys. *48*, 4803 (1968).
51) King, J., Elleman, D. D.: J. Chem. Phys. *48*, 412 (1968).
52) Mercea, V.: Isotopenpraxis *4*, 128 (1968).
53) Olah, G. A., Schleyer, P. R.: Carbonium ions. I. Interscience 1968.

H. Hartmann, K.-H. Lebert, and K.-P. Wanczek

54) Schaefer, J., Henis, J. M. S.: J. Chem. Phys. *49*, 5377 (1968).
55) Anders, L. R.: J. Chem. Phys. *73*, 4 9 (1969).
56) Beauchamp, J. L.: Tech. Progress Rep. Calif. Inst. Techn. Pasadena USA.
57) Beauchamp, J. L.: J. Am. Chem. Soc. *91*, 5925 (1969).
58) Beauchamp, J., Armstrong, J. T.: Rev. Sci. Instr. *40*, 123 (1969).
59) Bowers, M. T., Elleman, D. D.: 17th annual Conf. on Mass Spec. and Allied Topics, Dallas Texas USA 18.—23.5. (1969) ASTM Comm E—14. (P 381—2).
60) Clow, R. P., Futrell, J. H.: 17th annual Conf. on Mass. Spec. and Allied Topics, Dallas Texas USA 18.—23.5. (1969) ASTM Comm E—14. (P 394).
61) Huntress, W. T., Elleman, D. D.: 17th annual Conf. on Mass. Spec. and Allied Topics, Dallas Texas USA 18.—23.5. (1969) ASTM Comm E—14. (P 380).
62) Bowers, M. T., Elleman, D. D., King, J., Jr.: J. Chem. Phys. *50*, 4787 (1969).
63) Bowers, M. T., Elleman, D. D.: J. Chem. Phys. *51*, 4606 (1969).
64) Bowers, M. T., Elleman, D. D., King, J.: J. Chem. Phys. *50*, 1840 (1969).
65) Brauman, J. I., Blair, L. K.: J. Am. Chem. Soc. *91*, 2126 (1969).
66) Brauman, J. I., Smyth, K. C.: J. Am. Chem. Soc. *91*, 7778 (1969).
67) Buttrill, S. E.: J. Chem. Phys. *50*, 4125 (1969).
68) Clow, R. P., Futrell, J. H.: J. Chem. Phys. *50*, 5041 (1969).
69) Diekman, J., McLoed, J., Djerassi, C., Baldeschwieler, J.: J. Am. Chem. Soc. *91*, 2069 (1969).
70) Diekman, J. D.: Diss. Abstr. Order No *69*, 17408; *30*, 1587-B (1969).
71) Eadon, G., Diekman, J., Djerassi, C.: J. Am. Chem. Soc. *91*, 3986 (1969).
72) Henis, J. M. S.: Anal.Chem.*41*, 22A-26A (1969); 28A-30A (1969); 32A (1969).
73) Holz, J., Beauchamp, J. L.: J. Am. Chem. Soc. *91*, 5913 (1969).
74) Huntress, W. T., Baldeschwieler, J. D., Ponnamperuma, C.: Nature *223*, 468 (1969).
75) Huntress, W. T., Beauchamp, J. L.: Intern. J. Mass Spectr. Ion Phys. *3*, 149 (1969).
76) Inoue, M., Wexler, S.: J. Am. Chem. Soc. *91*, 5730 (1969).
77) Kitsenko, A. B.: Ukr. Fiz. Zh. *14*, 1515 (1969).
78) Llewellyn, P. M.: U.S.Pat. 3475605 Varian (1969).
79) O'Malley, R. M., Jennings, K. R.: Intern. J. Mass Spectr. Ion Phys. *2*, App. 1—3 (1969).
80) O'Malley, R. M., Jennings, K. R.: Intern. J. Mass Spectr. Ion Phys. *2*, 257 (1969).
81) O'Malley, R. M., Jennings, K. R.: Intern. J. Mass Spectr. Ion Phys. 2, 441 (1969).
82) Ridge, D. P., Beauchamp, J. L.: J. Chem. Phys. *51*, 470 (1969).
83) Schaefer, J., Henis, J. M. S.: Chem. Phys. *51*, 4671 (1969).
84) —.
85) Wilkins, C. L., Gross, M. L.: 5th Midwest Regional Meeting Am. Chem. Soc., Kansas City, Abstr. No. 405 (1969).
86) Aplin, R. T.: Ann. Rep. Prog. Chem. *67*, B 7—17 (1970).
87) Baldeschwieler, J. P.: US Pat. 3535512 (1970).
88) Beauchamp, J. L.: US Pat. 3 502867 (1970).
89) Beauchamp, J. L.: Vortex *31*, 24 (1970).
90) Beauchamp, J. L.: Nasa Star *8*, 3167 (1970).
91) Beauchamp, J. L., Dunbar, R. C.: J. Am. Chem. Soc. *92*, 1477 (1970).
91a) Benezra, S. A., Hoffman, M. K., Bursey, M. M.: J. Am. Chem. Soc. *92*, 7501 (1970).
92) Billets, S., Jaffé, H. H., Kaplan, F.: J. Am. Chem. Soc. *92*, 6964 (1970).
93) Bloom, M.: In: Atomic physics 2 (ed. P. G. H. Sanders). Plenum Press 1970.

110

94) Bowers, M. T., Elleman, D. D.: J. Am. Chem. Soc. *92*, 1847 (1970).
95) Bowers, M. T., Elleman, D. D.: J. Am. Chem. Soc. *92*, 7258 (1970).
96) —.
97) —.
98) —.
99) —.
100) Bowers, M. T., Elleman, D. D., O'Malley, R. M., Jennings, K. R.: J. Phys. Chem. *74*, 2583 (1970).
101) Brauman, J. I.: Chem. Inst. Canada and Amer. Chem. Soc. Joint Conf. Toronto 24—25. 5. 70, Abstr. NoOrgn. 77.
102) —.
103) Bursey, M. M., Elwood, T. A., Hoffman, M. K., Lehmann, T. A., Tesarek, J. M.: Anal. Chem. *42*, 1370 (1970).
104) Bursey, M. M., Benezra, S. A., Hoffman, M. K.: J. Am. Chem. Soc. *92*, 205 (1970).
105) Buttrill, S. E.: Diss. Abstr. Order No 70-18383 *31*, 1859-B 224 (1970).
106) Buttrill, S. E.: J. Am. Chem. Soc. *92*, 3560 (1970).
107) Clow, R. P., Futrell, J. H.: Intern. J. Mass Spectr. Ion Phys. *4*, 165 (1970).
108) Dunbar, R. C.: Diss. Abstr. Order No 70-22164 *31*, 2595 B (1970).
109) Dunbar, R. C.: J. Chem. Phys. *52*, 2780 (1970).
110) Eadon, G., Diekman, J., Djerassi, C.: J. Am. Chem. Soc. *92*, 6205 (1970).
111) Eyler, J. R.: Inorg. Chem. *9*, 981 (1970).
112) Fujiwara, S.: Magnetic resonance (ed. C. K. Coogan) Plenum Press 1970.
113) Gabowich, M. D., Soloshenko, I. A.: Soviet Phys.-Tech. Phys. (English Transl.) *15*, 184 (1970).
114) Goode, G. C., O'Malley, R. M., Ferrer-Correia, A. J., Massey, R. I., Jennings, K. R.: Intern. J. Mass Spectr. Ion Phys. *5*, 393, (1970).
115) Goode, G. C., Ferrer-Coreia, A. J., Jennings, K. R.: Intern. J. Mass Spectr. Ion Phys. *5*, 229 (1970).
116) Goode, G. C., O'Malley, R. M., Ferrer-Correia, A. J., Jennings, K. R.: Nature *227*, 1093 (1970).
117) Graeves, C.: Vacuum *20*, 65 (1970).
118) Henis, J. M. S.: J. Chem. Phys. *52*, 292 (1970).
119) Henis, J. M. S.: J. Chem. Phys. *52*, 282 (1970).
120) Henis, J. M. S., Mabie, C. A.: J. Chem. Phys. *53*, 2999 (1970).
121) Herod, A. A., Harrison, A. G., O'Malley, R. M., Ferrer-Correia, A. J., Jennings, K. R.: J. Phys. Chem. *74*, 2720 (1970).
122) Hoffman, M. K., Elwood, T. A., Lehmann, T. A., Bersey, M. M.: Tetrahedron Letters *46*, 4021 (1970).
123) Holtz, D., Beauchamp, J. L., Eyler, J. R.: J. Am. Chem. Soc. *92*, 7045 (1970).
124) Holtz, D., Beauchamp, J. L., Woodgate, S. P.: J. Am. Chem. Soc. *92*, 7484 (1970).
125) Huntress, W. T., Ellemann, D. D.: J. Am. Chem. Soc. *92*, 3565 (1970).
126) De Jong, D. C.: Anal. Chem. *42*, 196R (1970).
127) Kaplan, F., Cross, P., Prinstein, R.: J. Am. Chem. Soc. *92*, 1455 (1970).
128) Kriemler, P., Buttrill, S. E.: J. Am. Chem. Soc. *92*, 1123 (1970).
129) Billets, S., Jaffé, H. H.: Am. Chem. Soc. 159th National Meeting, Houston, Texas 1970.
130) Lebert, K. H.: Meßtechnik *78*, 109 (1970).
131) Lehmann, T. A., Elwood, T. A., Hoffman, M. K., Bursey, M. M.: J. Chem. Soc. *B 1970*, 1917.
132) Llewellyn, P. M.: U.S. Pat. Varian Ass. (1970).

111

H. Hartmann, K.-H. Lebert, and K.-P. Wanczek

132a) Llewellyn, P. M.: U.S. Pat. 3475605 Varian, Nucl. Sci. Abstr. *24*, 56N°515 (1970).
133) Llewellyn, P. M.: U.S. Pat. 3505517 Varian Ass (1970).
134) —.
135) Luxon, J. L., Rich, A.: Proc. Int. Conf. on Precision Meas. and Fundamental Const., Gaithersburg, Maryland, August 1970.
136) Marshall, A. G.: Diss. Abstr. Order No 70-18445, *31*, 1878 B 217 (1970).
137) Marshall, A. G., Buttrill, S. E.: J. Chem. Phys. *52*, 2752 (1970).
138) McIver, Jr., R. T.: Rev. Sci. Instr. *41*, 555 (1970).
139) McIver, R. T.: Rev. Sci. Instr. *41*, 126 (1970).
140) Mosesman, M., Huntress, W.: J. Chem. Phys. *53*, 462 (1970).
141) Nixon, W. B., Bursey, M. M.: Tetrahedron Letters *50*, 4389 (1970).
142) Ryan, K. R.: Rev. Pure Appl. Chem. *20*, 81 (1970).
143) Trajmar, S., Rice, J. K., Kuppermann, A.: Advan. Chem. Phys. *18*, 15 (1970).
144) Negative Ions: Variation on a theme, Nature [Phys. Sci.] *230*, 142 (1971).
145) Adler, M. S., Sesturia, S. D.: Rev. Sci. Instr. *42*, 704 (1971).
146) Baldeschwieler, J. D., Woodgate, S. S.: Acc. Chem. Res. *4*, 114 (1971).
147) Beauchamp, J. L.: Ann. Rev. Phys. Chem. *22*, 527 (1971).
147a) Bowers, M. T., Aue, D. H., Webb, H. M., McIver, R. T.: J. Am. Chem. Soc. *93*, 4314 (1971).
147b) Bowers, M. T., Kemper, P. R.: J. Am. Chem. Soc. *93*, 5352 (1971).
148) Brauman, J. I., Blair, L. K.: J. Am. Chem. Soc. *93*, 3911 (1971).
149) Brauman, J. I., Blair, L. K., Riveros, J. M.: J. Am. Chem. Soc. *93*, 3914 (1971).
149b) Brauman, J. I., Blair, L. K.: J. Am. Chem. Soc. *93*, 4315 (1971).
150) Bursey, M. M., Hoffman, M. K., Benezra, S. A.: J. Chem. Soc. D 1971, 1417.
151) Buttrill, S. E.: Proc. Ann. Conf. Mass Spectr. All. Topics 19th, Atlanta.
152) Comisarow, M. B.: J. Chem. Phys. *55*, 205 (1971).
153) Dunbar, R. C.: J. Am. Chem. Soc. *93*, 4354 (1971).
154) Dunbar, R. C.: J. Am. Chem. Soc. *93*, 4167 (1971).
155) Dunbar, R. C.: J. Chem. Phys. *54*, 711 (1971).
156) Elwood, T. A.: Diss. Abstr. 72—10715 *32*, 5072 B 2339 (1971).
157) Forster, M. S., Beauchamp, J. L.: J. Am. Chem. Soc. *93*, 4924 (1971).
158) Futrell, J. H.: ICR Mass spectrometry. In: Dynamic mass spectrometry (ed. D. Price), Vol. 2 (1971).
159) Futrell, J. H., Smith, D., Clow, R.: Advan. Mass Spectr. *5*, 202 (1971).
160) Goode, G. C., O'Malley, R. M., Ferrer-Correia, A. J., Jennings, K. R.: Advan. Mass Spectr. *5*, 195 (1971).
161) Goode, G. C., O'Malley, R. M., Ferrer-Correia, A. J., Jennings, K. R.: Chem. Brit. *7*, 12 (1971).
162) Gray, G. A.: Advan. Chem. Phys. *19*, 141 (1971).
163) Gross, M. L.: J. Am. Chem. Soc. *93*, 253 (1971).
164) Gross, M. L., Wilkins, C. L.: Anal. Chem. *43*, 1624 (1971).
165) Gross, M. L., Wilkins, C. L.: Anal. Chem. *43*, 65 A (1971).
166) Hoffman, M. K., Bursey, M. M.: Tetrahedron Letters *27*, 2539 (1971).
167) Holtz, D., Beauchamp, J. L.: Science *173*, (1971).
168) Holtz, D., Beauchamp, J. L.: Nature [Phys. Sci.] *231* 204 (1971).
169) Beauchamp, J. L., Holtz, D., Henderson, W. G., Taft, R. W.: Inorg. Chem. *10*, 201 (1971).
170) Huntress, Jr., W. T., Mosesman, M. M., Ellemann, D. D.: J. Chem. Phys. *54*, 843 (1971).
171) Huntress, Jr., W. T.: J. Chem. Phys. *55*, 164 (1971).
172) Jagarana, R.: Rev. J. Anal. Chem. USSR *26*, 2, 175 (1971).

112

[173] Lehman, T. A., Elwood, T. A., Bursey, J. T., Bursey, M. M., Beauchamp, J. L.: J. Am. Chem. Soc. *93*, 2108 (1971).
[174] Marshall, A. G.: J. Chem. Phys. *55*, 1343 (1971).
[175] 19th Annual Conf. on Mass Spectr. and Allied Topics, Atlanta Georgia, USA, 1971, Am. Soc. for Mass Spectr. Astma Committee E 14.
[180] McIver, R. T., Dunbar, R. C.: Intern. J. Mass Spectr. Ion Phys. *7*, 471 (1971).
[181] McLafferty, F. W., Gross, M. L.: J. Am. Chem. Soc. *93*, 1267 (1971).
[182] McMahon, T. B., Beauchamp, J. L.: Rev. Sci. Instr. *42*, 1632 (1971).
[183] Parker, J. E., Lehrle, R. S.: Intern. J. Mass Spectr. Ion Phys. *7*, 421 (1971).
[184] Ridge, D. P., Beauchamp, J. L.: J. Am. Chem. Soc. *93*, 5925 (1971).
[185] Rudolph, P. S., Baldock, R.: SRNL-4706 Chem. Div. Ann. Prog. Rept. for period ending 20.5.1971.
[186] Smyth, K. C., McIver, R. T., Brauman, J. I.: J. Chem. Phys. *54*, 2758 (1971).
[187] Su, T. C. K.: Diss. Abstr. No 72-14623, *32*, 6341 B (1971).
[188] Wanczek, K.-P., Lebert, K.-H., Hartmann, H.: Z. Naturforsch. *27a*, 155 (1971).
[189] Wexler, S., Pobo, L. G.: J. Am. Chem. Soc. *93*, 1327 (1971).
[190] Wilkens, C. L., Gross, M. L.: J. Am. Chem. Soc. *93*, 895 (1971).
[191] Williams, D. H.: Advan. Mass Spectr. *5*, 569 (1971).
[192] Arnett, E. M., Jones, F. M., Taagepera, M., Henderson, W. G., Beauchamp, J. L., Holtz, D., Taft, R. W.: J. Am. Chem. Soc. *94*, 4727 (1972).
[193] Beauchamp, J. L., Holtz, D., Woodgate, S. D., Patt, S. L.: J. Am. Chem. Soc. *94*, 2798 (1972).
[194] Beauchamp, J. L., Caserio, M. C.: J. Am. Chem. Soc. *94*, 2638 (1972).
[194a] Beauchamp, J. L.: Am. Chem. Soc. 163rd Mtg., Boston 1972.
[195] Benezra, S. A.: Diss. Abstr. 72-18378 *32*, 6877 B 1558 (1972).
[196] Benezra, S. A., Bursey, M. M.: J. Am. Chem. Soc. *94*, 1024 (1972).
[197] Clow, R. P., Futrell, J. H.: J. Am. Chem. Soc. *94*, 3749 (1972).
[198] Am. Chem. Soc. 163rd Nat. Mtg. Boston; 972, Abstr..
[199] —.
[200] Clow, R. P., Futrell, J. H.: Intern. J. Mass Spectr. Ion Phys. *8*, 119 (1972).
[201] Dixon, D. A., Holtz, D., Beauchamp, J. L.: Inorg. Chem. *11*, 960 (1972).
[202] Drewery, C. J., Goode, G. C., Jennings, K. R.: Mass spectrometry, Vol. 5, 183 (ed. A. Maccoll).
[203] Dunbar, R. C.: J. Phys. Chem. *76*, 2467 (1972).
[204] Dunbar, R. C., Shen, J., Olah, G. A.: J. Chem. Phys. *56*, 3794 (1972).
[205] Eadon, G. A.: Diss. Abstr. No 72-11537 *32*, 5687 B (1972).
[206] Foster, M. S., Beauchamp, J. L.: J. Am. Chem. Soc. *94*, 2425 (1972).
[207] Gross, M. L.: J. Am. Chem. Soc. *94*, 3744 (1972).
[208] Gross, M. L., Lin, P.-H., Franklin, S. J.: Anal. Chem. *46*, 974 (1972).
[209] Hasted, J. B.: Physics of atomic collisions. London: Butterworth 1972.
[210] Hayashi, K.: Nucl. Sci. Abstr. *26*, 2675 (1972).
[211] Hass, J. R., Bursey, M. M., Kingston, D. G. I.: J. Am. Chem. Soc. *94*, 5094 (1972).
[212] Henderson, W. G., Taagepera, M., Holtz, D., McIver, R. T., Beauchamp, J. L., Taft, R. W.: J. Am. Chem. Soc. *94*, 4728 (1972).
[213] Henis, J. M. S., Stewart, G. W., Tripodi, M. K., Gaspar, P. P.: J. Chem. Phys. *57*, 389 (1972).
[214] Huntress, W. T.: J. Chem. Phys. *56*, 5111 (1972).
[215] Jaffé, H. H., Billets, S.: J. Am. Chem. Soc. *94*, 6745 (1972).
[216] Kevan, L., Futrell, J. H.: J. Chem. Soc. Faraday II *68*, 1742 (1972).
[217] Kramer, J. M., Dunbar, R. C.: J. Am. Chem. Soc. *94*, 4346 (1972).
[218] Lieder, C. A., Wien, R. W., McIver, R. T.: J. Chem. Phys. *56*, 5184 (1972).

H. Hartmann, K.-H. Lebert, and K.-P. Wanczek

219) Liedtke, R. J., Gerrard, A. F., Diekman, J., Djerassi, C.: J. Org. Chem. *37*, 776 (1972).
220) Mac Neil, K. A. G., Futrell, J. H.: J. Phys. Chem. *76*, 409 (1972).
221) Mc Adoo, D. J., Mc Lafferty, F. W., Bente, P. F.: J. Am. Chem. Soc. *94*, 2027 (1972).
222) Mc Allister, T.: J. Chem. Soc. Chem. Commun. 1972, 1245.
223) Mc Allister, T.: Intern. J. Mass Spectr. Ion Phys. *9*, 127 (1972).
224) Mc Allister, T.: J. Chem. Phys. *56*, 5192 (1972).
225) Mc Allister, T.: Chem. Phys. Letters *13*, 602 (1972).
226) Mc Allister, T.: Intern. J. Mass Spectr. Ion Phys. *8*, 162 (1972).
227) McMahon, T. B., Beauchamp, J. L.: Rev. Sci. Instr. *43*, 509 (1972).
228) Morton, T. H., Beauchamp, J. L.: Am. Chem. Soc. 163rd Mtg. Boston, Mass. 1972, Abstr. N. Orgn. 78 (1972).
229) Morton, T. H., Beauchamp, J. L.: J. Am. Chem. Soc. *94*, 3672 (1972).
229a) Mc Neil, K. A. G., Futrell, J. H.: J. Phys. Chem. *76*, 409 (1972).
230) Pasechnik, L. L., Prokhorov, V. V.: Yagola, Soviet Phys. Tech. Phys. *17*, 220 1970.
231) Sharp, T. E., Eyler, J. R., Li, E.: Intern. J. Mass Spectr. Ion Phys. *9*, 421, (1972).
232) Smith, D. H.: Mass spectrometry (ed. T. H. Gouw), p. 351. New York: Wiley 1972.
233) Smyth, K. C., Brauman, J. I.: J. Chem. Phys. *56*, 1132 (1972).
234) Smyth, K. C., Brauman, J. I.: J. Chem. Phys. *56*, 4620 (1972).
235) Smyth, K. C., Brauman, J. I.: J. Chem. Phys. *56*, 5993 (1972).
236) Stewart, G. W., Henis, J. M. S., Gaspar, P. P.: J. Chem. Phys. *57*, 1990 (1972).
237) Stewart, G. W., Henis, J. M. S., Gaspar, P. P.: J. Chem. Phys. *57*, 2247 (1972).
237a) Taagepera, M., Holtz, D., Mc Iver, R. T., Beauchamp, J. L., Taft, R. W.: J. Am. Chem. Soc. *94*, 4728 (1972).
238) Taagepera, M., Henderson, W. G., Brownlee, R. T. C., Holtz, D., Beauchamp, J. L.: J. Am. Chem. Soc. *94*, 1369 (1972).
239) Thean, J. E., Johnsen, R. H.: Intern. J. Mass Spectr. Ion Phys. *9*, 498 (1972).
240) Tse, R. S.: Intern. J. Mass Spectr. Ion Phys. *9*, 351 (1972).
241) Williams, D. H.: In: Proc. 2nd Conf. on Mass Spectr., Ispra, Italy (1971) Ed. Euratom, CID Luxemburg, 1972.
242) Wilson, M. H., Mc Closkey, J. A.: J. Am. Chem. Soc. *94*, 3865 (1972).
243) Author's unpublished results.
244) Anicich, V. G., Bowers, M. T., O'Malley, R. M., Jennings, K. R.: Intern. J. Mass Spectr. Ion Phys. *11*, 99 (1973).
245) O'Malley, R. M., Jennings, K. R., Bowers, M. T., Anicich, V.: Intern. J. Mass Spectr. Ion Phys. *11*, 89 (1973).
246) Blint, R. J., McMahon, T. B., Beauchamp, J. L.: private communication.
247) Dawson, J. H. J., Henderson, W. G., O'Malley, R. M., Jennings, K. R.: Intern. J. Mass Spectr. Ion Phys. *11*, 61 (1973).
248) Dunbar, R. C., Kramer, J. M.: Acc. J. Chem. Phys. 1973.
249) Dunbar, R. C.: Acc. J. Am. Chem. Soc. 1973.
250) Dunbar, R. C., Shen, J., Olah, G. A.: Acc. J. Am. Chem. Soc. 1973.
251) Dymerski, P. P., Dunbar, R. C.: J. Chem. Phys. 1973.
252) Ferrer-Correia, A. J., Jennings, K. R.: Intern. J. Mass Spectr. Ion Phys. *11*, 111 (1973).
253) Futrell, J. H.: private communication.
254) Gauglhofer, J., Kevan, L.: Chem. Phys. Letters
255) Gross, M. L.: Org. Mass Spectr. 6

256) Klopman, G., ed.: Wiley Interscience: R. C. Dunbar.
257) Marx, R., Mauclaire, G.: Intern. J. Mass Spectr. Ion Phys. *10*, 213 (1972/73).
258) Marx, R., Mauclaire, G., Fehsenfeld, F. C., Dunkin, D. B., Ferguson, E. E.: J. Chem. Phys. 58, 3267 (1973).
259) McAllister, T.: Intern. J. Mass Spectr. Ion Phys. *10*, 419 (1972/73).
260) McMahon, T. B., Blint, R. J., Ridge, D. P., Beauchamp, J. L.: J. Am. Chem. Soc. 1973.
261) Pobo, L. G., Wexler, S.: Radiochim. Acta 1973.
262) Smith, D. L., Futrell, J. H.: Intern. J. Mass Spectr. Ion Phys. *10*, 405 (1972/73).
263) Smith, D. L., Futrell, J. H.: private communication.
264) Wayatt, R. H., Holtz, D., McMahon, T. B., Beauchamp, J. L.: 1973.

Received March 26, 1973

Structure and Bonding

Editors: P. Hemmerich, C. K. Jørgensen, J. B. Neilands,
Sir Ronald S. Nyholm, D. Reinen, R. J. P. Williams

Volume 10

Inorganic Chemistry

49 figures. III, 190 pages. 1972
DM 58,–; US $ 26.20
ISBN 3-540-05700-5

Volume 11

58 figures. III, 170 pages. 1972
DM 54,–; US $ 24.40
ISBN 3-540-05830-3

Contents: Thomson, A. J.; Williams,
R. J. P.; Reslova, S., The Chemistry of
Complexes Related to cis-$Pt(NH_3)_2Cl_2$.
An Anti-Tumour Drug. – Wood, J. M.;
Brown, D. G., The Chemistry of
Vitamin B_{12}-Enzymes. – Bray, R. C.;
Swann, J. C., Molybdenum-Containing
Enzymes. – Neilands, J. B., Evolution
of Biological Iron Binding Centers.

Volume 12

Progress in Theory

37 figures. III, 295 pages. 1972
DM 72,–; US $ 32.50
ISBN 3-540-05901-6

Volume 13

70 figures. III, 253 pages. 1973
DM 72,–; US $ 32.50
ISBN 3-540-06125-8

Contents: Penneman, R. A.; Ryan,
R. R.; Rosenzweig, A., Structural
Systematics in Actinide Fluoride
Complexes. – Reisfeld, R., Spectra and
Energy Transfer of Rare Earths in
Inorganic Glasses. – Felsche, J., The
Crystal Chemistry of the Rare – Earth
Silicates. – Jørgensen, C. K., The Inner
Mechanism of Rare Earths Elucidated
by Photo-Electron Spectra.

Volume 14

52 figures. III, 172 pages. 1973
DM 56,–; US $ 25.30
ISBN 3-540-06162-2

With contributions by: A. Ludi,
H. U. Güdel, A. Müller, E. Diemann,
C. K. Jørgensen, B. J. Hathaway,
C. E. Schäffer, B. Magyar, U. Müller

Volume 15

Coordinative Inter-actions

Approx. 70 figures. Approx. 160 pages.
1973. In preparation
ISBN 3-540-06410-9

Prices are subject to change without notice

Springer-Verlag
Berlin
Heidelberg
New York
München London Paris
Sydney Tokyo Wien

NMR

Basic Principles and Progress
Grundlagen und Fortschritte
Editors:
P. Diehl, E. Fluck, R. Kosfeld

Prices are subject to change without notice

Springer-Verlag
Berlin
Heidelberg
New York
München · London · Paris
Sydney · Tokyo · Wien